ENERGY 2018

The Eighth International Conference on Smart Grids, Green Communications and IT Energy-aware Technologies

May 20 - 24, 2018

Nice, France

ENERGY 2018 Editors

Michael Negnevitsky, University of Tasmania, Australia
Rafael Mayo-García, CIEMAT, Spain
José Mª Cela, BSC-CNS, Spain
Alvaro LGA Coutinho, COPPE-UFRJ, Brazil
Vivian Sultan, Center of Information Systems and Technology, Claremont Graduate University, USA

Printed from e-media with permission by:

Curran Associates, Inc.
57 Morehouse Lane
Red Hook, NY 12571

Some format issues inherent in the e-media version may also appear in this print version.

Printed by Curran Associates, Inc. (2018)

International Academy, Research, and Industry Association (IARIA)
412 Derby Way
Wilmington, DE 19810

Phone: (408) 893-6407
Fax: (408) 527-6351

petre@iaria.org

Additional copies of this publication are available from:

Curran Associates, Inc.
57 Morehouse Lane
Red Hook, NY 12571 USA
Phone: 845-758-0400
Fax: 845-758-2633
Email: curran@proceedings.com
Web: www.proceedings.com

Table of Contents

ENERGY 2018

Foreword

The Eighth International Conference on Smart Grids, Green Communications and IT Energy-aware Technologies (ENERGY 2018), held between May 20 - 24, 2018 - Nice, France, continued the event considering Green approaches for Smart Grids and IT-aware technologies. It addressed fundamentals, technologies, hardware and software needed support, and applications and challenges.

There is a perceived need for a fundamental transformation in IP communications, energy-aware technologies and the way all energy sources are integrated. This is accelerated by the complexity of smart devices, the need for special interfaces for an easy and remote access, and the new achievements in energy production. Smart Grid technologies promote ways to enhance efficiency and reliability of the electric grid, while addressing increasing demand and incorporating more renewable and distributed electricity generation. The adoption of data centers, penetration of new energy resources, large dissemination of smart sensing and control devices, including smart home, and new vehicular energy approaches demand a new position for distributed communications, energy storage, and integration of various sources of energy.

We take here the opportunity to warmly thank all the members of the ENERGY 2018 Technical Program Committee, as well as the numerous reviewers. The creation of such a high quality conference program would not have been possible without their involvement. We also kindly thank all the authors who dedicated much of their time and efforts to contribute to ENERGY 2018. We truly believe that, thanks to all these efforts, the final conference program consisted of top quality contributions.

Also, this event could not have been a reality without the support of many individuals, organizations, and sponsors. We are grateful to the members of the ENERGY 2018 organizing committee for their help in handling the logistics and for their work to make this professional meeting a success.

We hope that ENERGY 2018 was a successful international forum for the exchange of ideas and results between academia and industry and for the promotion of progress in the fields of smart grids, green communications and IT energy-aware technologies.

We are convinced that the participants found the event useful and communications very open. We also hope that Nice provided a pleasant environment during the conference and everyone saved some time for exploring this beautiful city.

ENERGY 2018 Chairs:

ENERGY Steering Committee

Steffen Fries, Siemens, Germany
Akhtar Kalam, Victoria University, Australia
Vincenzo Gulisano, Chalmers University of Technology, Sweden
Marco Lützenberger, Technische Universität Berlin, Germany
Evangelos Pournaras, ETH Zurich, Switzerland
Angela Russo, Politecnico di Torino, Italy
Chun-Hsi Huang, University of Connecticut, USA
Mark Apperley, University of Waikato, New Zealand
Magnus Almgren, Chalmers University of Technology, Sweden

ENERGY 2018

Committee

ENERGY Steering Committee
Steffen Fries, Siemens, Germany
Akhtar Kalam, Victoria University, Australia
Vincenzo Gulisano, Chalmers University of Technology, Sweden
Marco Lützenberger, Technische Universität Berlin, Germany
Evangelos Pournaras, ETH Zurich, Switzerland
Angela Russo, Politecnico di Torino, Italy
Chun-Hsi Huang, University of Connecticut, USA
Mark Apperley, University of Waikato, New Zealand
Magnus Almgren, Chalmers University of Technology, Sweden

ENERGY Industry/Research Advisory Committee
Hongbo Sun, Mitsubishi Electric Research Laboratories, USA
Daisuke Mashima, Advanced Digital Sciences Center, Singapore
Sambaran Bandyopadhyay, IBM Research, India
Eric MSP Veith, OFFIS e.V. - Oldenburg, Germany
Dragan Obradovic, Siemens - Corporate Technology, Munich, Germany
Chresten Træholt, Center for Electric Power and Energy - CEE | Technical University of Denmark - DTU, Denmark

ENERGY 2018 Technical Program Committee

Cristinel Ababei, Marquette University, USA
Kouzou Abdellah, Djelfa University, Algeria
Hatim G. Abood, The University of Western Australia, Australia
A. Abu-Siada, Curtin University, Australia
Baris Aksanli, San Diego State University, USA
Ahmed Al-Salaymeh, University of Jordan, Jordan
Magnus Almgren, Chalmers University of Technology, Sweden
Mark Apperley, University of Waikato, New Zealand
Sambaran Bandyopadhyay, IBM Research, India
Chakib Bekara, University of Tlemcen, Algeria
Rachid Benchrifa, Mohammed V University, Morocco
Andrea Bonfiglio, University of Genoa, Italy
Massimo Brignone, University of Genoa, Italy
Concettina Buccella, University of L'Aquila, Italy
Erik Buchmann, Hochschule für Telekommunikation Leipzig (University of Applied Sciences) Leipzig, Germany
Vito Calderaro, University of Salerno, Italy
Davide Careglio, Universitat Politècnica de Catalunya, Spain
Hsi-Ya Chang, National Center for High-performance Computing, Taiwan
Antonin Chazalet, Orange, France

Ali Cheknane, Amar Telidji University of Laghouat. Algeria
Luigi Costanzo, Università degli Studi della Campania Luigi Vanvitelli, Italy
Narottam Das, University of Southern Queensland Toowoomba, Australia
Margot Deruyck, iMinds - Ghent University - INTEC - WAVES, Belgium
Vincenzo Di Dio, University of Palermo, Italy
Efrén Díez Jiménez, Universidad de Alcalá, Spain
Rekioua Djamila, University of Bejaia, Algeria
Khaled Elbassioni, Masdar Institute of Science and Technology, Abu Dhabi, UAE
Rolf Egert, Technische Universität Darmstadt, Germany
Tullio Facchinetti, University of Pavia, Italy
Daniel Freund, Distributed Artificial Intelligence Lab (DAI-Lab), Berlin, Germany
Steffen Fries, Siemens, Germany
Vincenzo Gulisano, Chalmers University of Technology, Sweden
Chun-Hsi Huang, University of Connecticut, USA
Zhiyi Huang, University of Otago, New Zealand
Dan M. Ionel, University of Kentucky, USA
Philip Johnson, University of Hawaii, USA
Akhtar Kalam, Victoria University, Australia
György Kálmán, Norwegian University of Science and Technology (NTNU)/mnemonic AS, Research Group on Critical Infrastructure Protection, Norway
Essam E. Khalil, Cairo University, Egypt
Tobias Küster, Technische Universität Berlin (TU Berlin), Germany
Elias Kyriakides, KIOS Research Center | University of Cyprus, Cyprus
Zhenhua Liu, Stony Brook University, USA
Abraham Lomi, National Institute of Technology, Malang, Indonesia
Marco Lützenberger, Technische Universität Berlin, Germany
Thair Mahmoud, Edith Cowan University, Australia
Apel Mahmud, Deakin University, Australia
Daisuke Mashima, Advanced Digital Sciences Center, Singapore
Rafael Mayo-García, CIEMAT, Spain
Marilena Minou, Athens University of Economics & Business, Greece
Hemanta Kumar Mondal, CNRS Lab-STICC - UMR 6285 | Université Bretagne Sud (UBS), France
Hugo Morais, ISEP/IPP, Portugal
Fabio Mottola, University of Naples Federico II, Italy
Gero Mühl, Universitaet Rostock, Germany
Masayuki Murata, Osaka University Suita, Japan
Md Mustafizur Rahman, Universiti Malaysia Pahang, Malaysia
Michael Negnevitsky, University of Tasmania, Australia
Rajiv Nishtala, Barcelona Supercomputing Center, Spain
Bruce Nordman, LBNL, USA
Dragan Obradovic, Siemens - Corporate Technology, Munich, Germany
Marina Papatriantafilou, Chalmers University of Technology, Sweden
Massimo Poncino, Politecnico di Torino, Italy
Evangelos Pournaras, ETH Zurich, Switzerland
Manuel Prieto-Matias, Complutense University of Madrid, Spain
Renato Procopio, University of Genoa, Italy
Jan Richling, South Westphalia University of Applied Sciences, Germany
Angela Russo, Politecnico di Torino, Italy

Mischa Schmidt, NEC Laboratories Europe, Heidelberg, Germany
Kenneth Schmitz, German Research Center for Artificial Intelligence - DFKI GmbH,Germany
Harald Schrom, Institut für Datentechnik und Kommunikationsnetze - TU Braunschweig, Germany
Taha Selim Ustun, Carnegie Mellon University, USA
Farhad Shahnia, Murdoch University, Perth, Australia
Vivian Sultan, Claremont Graduate University, USA
Hongbo Sun, Mitsubishi Electric Research Laboratories, USA
Chresten Træholt, Center for Electric Power and Energy - CEE | Technical University of Denmark - DTU, Denmark
Eric MSP Veith, OFFIS e.V. - Oldenburg, Germany
Massimo Vitelli, Università degli Studi della Campania Luigi Vanvitelli, Italy
Matthias Vodel, Technische Universitaet Chemnitz, Germany
Florian Volk, Technische Universität Darmstadt, Germany
Marcus Voß, Technische Universität Berlin (DAI-Labor), Germany

Copyright Information

The Future Development of Smart Grid

The case of Morocco

Nada Belkebir[1,2], Mohammed Maaroufi
[1] Ecole Mohammadia d'Ingénieurs
Mohammed V University
Rabat, Morocco
E-mail: nada.belkebir@gmail.com
maaroufi@emi.ac.ma

Ahmed Khallaayoun, Rachid Lghoul
School of Sciences and Engineering
[2] Al Akhawayn University in Ifrane
Ifrane, Morocco
E-mail: a.khallaayoun@aui.ma
r.lghoul@aui.ma

***Abstract*— Morocco is becoming more concerned about the protection of environment as it conducted different studies and strategies with the High Planning Commission (Haut Commissariat au Plan) and under the directives of top Moroccan managers whose purpose is to integrate sustainable energy in its economic activities. It is in this direction that Morocco prepared one of the largest studies "Prospective Maroc 2030" and in which a whole section was devoted to shine the light on the issues and challenges related to energy that the country has to face. With an expected growth of 5 to 6% in electricity consumption and with the integration of renewable energies, Smart Grid seems to be the natural transition to deal with the different fluctuations that the grid will be subjected to. This paper evaluates the current status of Smart Grid in Morocco in comparison with other developing countries. An overview of the energy sector in Morocco is presented. The regulations related to the power sector are also discussed. Then challenges and recommendations will be discussed at the end of the paper.**

Keywords- Smart Grid; Morocco; Renewable energies; Energy; Electricity

I. INTRODUCTION

As the electric grid got bigger and bigger, control and monitoring became imperative to ensure its management. The aging infrastructure and the ever increasing demand render the conventional grid susceptible to recurring failures. With the revolution of communication systems and the need to make the grid smarter, the idea of the Smart Grid (SG) saw light to enhance the functionality of the electric grid. The latter is achieved by introducing sensors, controls, and communication tools to enable a better management of energy. The electrical grid is divided into three major parts: generation, transmission and distribution. Today's generation relies on two different types of resources; fossil fuels, such as coal, oil or natural gas and renewable energies, such as wind, solar, biomass or hydropower. The inclusion onto the grid of the second type of energies makes it more difficult to manage due to the inherent intermittence of those resources. As for the transmission part, the electric grid suffers from losses which if managed correctly can save a significant amount of energy. As far as the distribution is concerned, electricity was conventionally distributed from retailers to customers. Nowadays, however, a new distribution concept was introduced, which consists of distributed generation. It consists of having small production units that the customer can own and can locally produce the electricity needed. SG aims then at assuring reliability, minimizing cost and managing resources [16].

Section II describes the features, the components and the challenges facing the SG. Section III presents the situation of SG in different countries. Section IV describes the current situation of the Moroccan electric grid and how electricy is governed. In addition, the government initiatives that aim at implementing the SG in Morocco and the challenges that the country is facing are included. Section V presents a comparison between Morocco and other countries when it come to SG.

II. DESCRIPTION OF SMART GRID

SG is a modern electric grid characterized by the use of sensors, communication technology and controls with the goal of applying intelligence to optimize generation, transmission, and distribution [15]. This section presents the main features of a typical SG and presents its main components.

A. SG Features

One of major features that characterizes a SG is its ability to manage a complex network that contains central and distributed generation that use both, conventional and renewable resources. The network uses a multitude of telecommunication schemes where internet and electronic applications are used to optimize the use of energy. These features create a new energy value chain capable of mixing different technologies and in which distributed generation is encouraged. With the introduction of small and medium power farms, the introduction of technology enables customers to manage their energy depending on the local production. Energy distribution is then managed depending on the different regulations set by the retailers or the customers. With the introduction of smart meters each customer has the ability to get a closer look at its consumption and manage it depending on the dynamic energy pricing which can therefore reduce the peak demand and flatten the demand curve [2]. The SG does not only manage energy at the home/building level, but it expends to managing whole districts by creating systems that combine data coming from different energy management systems. In other words, the aim of the SG is to manage the energy produced from micro-grids and the energy coming from macro stations in the most optimized manner, enabling a reduction in energy losses and minimization of energy cost. Mainly, the SG aims at reducing Green House Gases and CO_2 emissions, providing a better transmission and a better distribution [3].

B. SG Components

The SG is a set of individual technologies that cover the entire grid. In order to optimize the electric grid, different technologies continue to be deployed. The most important components of SG are the listed below [2]:

- *Wide-area monitoring and control*: concerns the generation and transmission side of the grid, it enables the grid to mitigate wide-area disturbances, and improve transmission capacity and reliability.
- *Information and communication technology*: spans the whole grid and is considered the backbone of the system. This technology starts from the generation level and goes down to the residential, industrial and service level. It can use different forms of communication networks that aim at providing the necessary support for data transmission. The communication technology used enable two way communication within the grid.
- *Renewable and distributed generation integration:* renewable energies need to be integrated at the transmission level and at different scales (High, medium, and low voltage). With the inherent intermittence of renewables, management of the various resources becomes more challenging. The introduction of storage and additional spinning reserves may then be required to ensure grid stability.
- *Advanced Metering Infrastructure (AMI)*: considered as the building blocks of the SG, smart meters allow customers to receive information about their energy consumption and inform them about other details, such as electricity prices, amount of energy consumed, losses and theft detection; which can help them better manage their electricity use and ultimately reduce their bills.
- *Electric vehicle charging infrastructure*: As the electric cars are introduced into the market, the load curve, due to the additional stress on the grid, might see a significant variation that might render the grid unstable. The advanced metering infrastructure will become the mean by which those variations are controlled.
- *Customer-side systems*: it concerns all systems that can be used either in the industrial or residential sectors that allow a better energy management and that can help increase energy efficiency. Those systems are energy management systems, energy storage devices, smart appliances and distributed generation; they connect the customer side to all the previous cited technologies that make the SG.

C. SG Communication and security

As mentioned before, the SG needs different components to get the intelligence necessary to optimize energy consumption and reduce the customer's bill. The grid is divided into two main parts. The first one concerns the high voltage lines that connect the power plant to the distribution substations. Electricity is then transformed via the substation to medium voltages after which it is distributed to building feeders. Communication between the power plant and the transmission substations is established using high speed optical fibers. The reason behind this choice is the need to transfer high volume of SG data, with the least possible delay. The second part includes different networks. The first network is the Neighborhood Area Network (NAN) that usually manages energy at the neighborhood level. This latter serves different Building Area Networks (BANs). Each BAN manages energy of each separate building, which means that it includes a number of Home Area Networks (HANs). Each HAN gathers a number of home appliances all connected to the HAN gateway. Communication at the HAN is achieved using ZigBee, since it requires a low power compared to Wi-Fi or Bluetooth connections. As mentioned before, the BAN gateway monitors the energy consumption of different HANs. Wi-Fi can be used as a communication tool here, but when the BAN covers many households in a large area it requires the deployment of additional access points. Having multiple access points in close proximity will interfere with other access point dedicated for internet usage. An alternative can be the use of 4G technologies, which can facilitate the communication between a BAN and its HANs. When it comes to the NAN, the connection can be done using 4G technologies as well. Figure 1 presents how communication is set between different networks discussed before [16].

It is imperative that the SG is secure and has the ability to prevent unwanted data exchange and malicious threats. For that purpose, a robust encryption mechanism needs to be implemented although it might cause a significant delay in data transmission.

D. SG Challenges

The most important challenge that the SG faces is to match supply and demand. In fact, the SG needs to adjust its generation depending on the data provided from the demand side. This latter is provided by different types of demand responses which require accurate forecast and customer participation. Privacy becomes an issue when customer consumption behavior is recorded; the SG needs to protect data coming from customers to avoid any privacy issues. Another challenge of the SG is to find a balance between minimizing the bill for the customer vs. maximizing the profit for the utility. Different types of contracts can be set between the two in order to reach an equilibrium depending on each specific case.

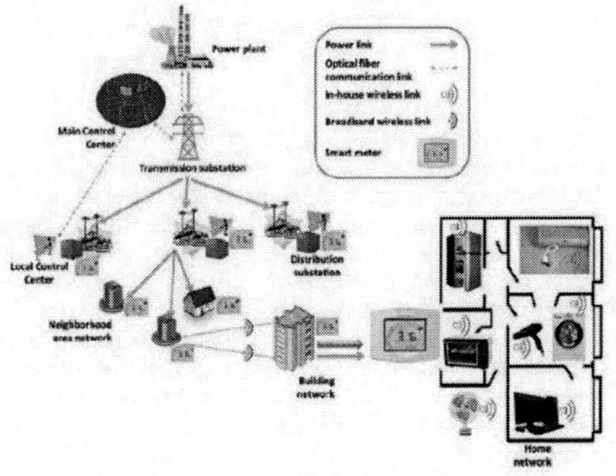

Figure 1. Hierarchical communication networks in a SG model [16]

Another challenge facing the SG concerns the transmission losses and the power loss due to storage. New algorithms need to be developed in order to optimize power distribution and transmission of electricity. Last but not least, security discussed in the previous part needs to be provided in order to prevent threats that might cause damage to the grid [16].

III. SG in Different Countries

After looking at the components and challenges of the SG, Section III will present what some countries have undertaken to develop a SG.

A. *SG in India*

India faces many challenges when it comes to the electric grid. Even though it has a good energy mix that includes conventional and renewable energies, the future of its economy is considered as being at risk if nothing is done regarding the quality of its electric grid. The total installed capacity of the country was 329 GW by the end of September 2017; 31.3% of its demand is met by renewable energy resources and it is expected to keep increasing by 2030-2040 in order to meet the demand [20]. As of 2014, only 79% of the population had access to electricity [19]. Moreover, the transmission and distribution losses at the end of the same year were estimated to 19% [19]. Because of the aforementioned challenges, the Central Electric Authority created the "National Smart Grid Mission (NSGM)". This new organization was created on March 27th, 2015 in order to monitor and implement new policies and programs related to the SG [20]. The Ministry of Power also released in 2013 a SG roadmap for India, in which the vision and the targets of the country when it comes to its electric grid were stated. The SG vision of India is to change the Indian electric grid into a safe and sustainable one, in which electricity is assured for all [21]. In order to achieve this vision, many sub-goals need to be achieved:

- At the distribution level: new policies need to be launched in order to guarantee access to electricity to all households by 2017. Another target of the NSGM is to reduce transmission and distribution losses to below 15% by 2017 and below 10% by 2027. After the success of the pilot projects of AMI in 2016, the Indian government is planning to develop smart meters with the necessary IT and communication infrastructure by 2027. Moreover, the development of Microgrids is also one of the main targets of NSGM; in fact, it is expected to have 10,000 Microgrids in villages, industrial parks or commercial hubs in order to reduce the peak demand by 2022 [21].
- At the transmission level: the NSGM intend to implement a Wide Area Monitoring Systems using Phasor Measurement Units for the whole transmission system by 2017. 50,000 Kms of optical fiber cables are planned to be installed in order to facilitate the implementation of the SG technologies. The Indian government plans also to set up a Renewable Energy Monitoring Center in order to ease the integration of renewables [21].

Releasing new policies and standards for all these initiatives is part of the SG roadmap, as they will enable an active involvment of customer. In fact, policies need to include Demand Response (DR) programs and dynamic pricing that can be used once the smart meters are installed [21]. It can be concluded that India is now aware of the importance of the SG for the future of its economy. The development of smart metering is the first step that will enable the country to get a better view on its energy consumption which will enable a better introduction of renewable energies to the Indian grid.

B. *SG in Turkey*

The electricity market of Turkey was for years controlled by the state, which created a high level of centralization [5]. By the 1980s generation, transmission and distribution started to be delegated to different companies. Small power plants started to see the light, which created a more efficient environment in which cheap energy generation was available. In 2004, new laws regarding how the privatization of the electric sector were published. All the transmission operations are led by the state control, while distribution is covered by private companies that are each responsible for their respective regions (as Turkey is divided into 21 regions) [6]. As for the distribution, private companies made some improvements in terms of the monitoring and control of electricity as they implemented the Supervisory Control and Data Acquisition (SCADA) software. This technology enables companies to get the necessary data to avoid any outages. Geographical Information Systems (GIS) is also an important tool used by the Turkish distribution companies as it enables locating places of high and low voltage consumers. Automated Meter Reading (AMR) is also used to collect data about the consumers' consumption. All data collected by the cited technologies is then integrated in a system to enable better optimization of electric energy in Turkey [7]. The total installed capacity of Turkey reached 80 GW in July 2017; 35% of this capacity is from renewable energy resources [22]. In 2014, the country was still suffering from 15% of electric power transmission and distribution losses [19]. Introducing smart metering will help reduce this percentage and increase energy supply efficiency; Turkey has dedicated $5 billion to invest in SG technologies in 2015 [5]. On the other hand, Turkey is also concerned with the introduction of renewable energies, as it launched a strategic plan between 2015 and 2019 that aims at providing energy security; by increasing the percentage by which renewable energies contribute to the energy supply and by keeping investments on transmission and distribution components as they are the connecting bridge between supply and demand [22]. Support programs for R&D in the SG domain are also very encouraged in the country [5].

C. SG in the U. S.

The total electric generation capacity of the U.S. reached 1,190.5 GW in 2017 with 17% of it coming from renewable energy resources [24]. The transmission and distribution losses of the U.S. was estimated to 6% in 2014 [19]. The modernization of the U.S. electric grid started in the 1980s. In 2004 the term "Smart Grid" got its first references by Amin and Wollenberg [7]. New policies and regulations that encourages the introduction of renewable energies and distributed generation were launched which led to significant investments by the American Recovery and Reinvestment Act of 2009 that reached $9 billion in 2010 [8]. Progress followed throughout the years in order to improve the many areas of the SG. In 2014, a total of 65 million smart meters were installed in the country; this number represents more than the third of electricity customers. The U.S. also intends to develop customer-based technologies, such as building energy management systems for residential, commercial and industrial customers. Those systems, and with the integration of sensors and AMI, can provide users with a better management of their energy which can increase efficiency in different sectors [8]. The U.S. SG Initiative launched in 2007 was a formalized way to set all the objectives the nation has to achieve, including the development of new storage and peak-shaving technologies, with a pulg-in hybrid electric vehicle [7]. In 2011, the annual SG spendings hit $5.2 billion. These investments were dedicated to the advanced SG projects, distribution automation and smart metering. It was expected that in 2017, the total investments in SG would be about $3.3 billion. The number may seem to decline, because less spending are dedicated to smart metering, but in reality it will gradually increase for the case of the distribution automation as it was $1.2 billion in 2011 and was expected to be $1.9 billion in 2017 [7]. Moreover, the U.S. intend to increase its energy mix by creating more solar, wind and nuclear plants, as they account for more than 10% of the pending plants application. As for the planned power plants, wind account for one third of the proposed plants [24].

D. SG in China

The Chinese power grid China was considered as being vulnerable. This conclusion was made after the snow storm of 2008 that caused a blackout in a major area of the country. Moreover, the power distribution of China's grid had a lack of intelligence which led to either seasonal shortages or surplus in different areas. The main objective of China is to have a strong grid that can enable a better integration of renewable energies and increase energy efficiency in order to adapt to climate change. On the other hand, the country needs to attract more investors in order to have a better privatization of its electric industry and therefore have more competitiveness in the market. Providing a safe power sector is ultimate if the country wants to have more investors [9]. For these purposes, China focuses on the development of its SG on high voltage and main network intelligence. In other words, the developpement will focus on increasing the transmission capacity and security of energy supply [10]. The SG concept in China is a bit different from the one in western countries. As a matter of fact, China intends to invest largely in large power plants; and for that purpose it is building a strong Ultra-High-Voltage (UHV) power network in order to keep increasing transmission efficiency and reduce energy losses. It is in May 2009 that the State Grid Corporation of China (SGCC) shared it's strategic plan to make the chinese grid smart. Considered as one of the priorities of the chinese government, the implementation of the SG in China follows three part [27]:

- 2009-2010: it is during this period that pilot projects were created in order to test what are the best options for the chinese grid and make the master plan for he SG
- 2011-2015: called the construction period, it is during this period that China implemented the main SG technologies that can help keep up with the extensive growth of the energy sector of the country
- 2016-2020: it is the final period where all the upgrades should be made in order to acheive a reliable and strong grid, in which China can optimize its grid at all levels

With its 1,645.75 GW of total installed capacity recorded in 2016 [26], China considerably reduced its transmission and distribution losses as they reached 5.4% in 2014 [19]. The chinese government is also aware of the importance of the energy mix. In 2016, 24.8% of the total electricity of the country was generated from renewable energy resources. This percentage is expected to reach 30% by 2020 [27]. To meet the demand, China is developping a smart dispatch and control technologies that can make energy available countrywide. The last characteristic of the chineese SG is that customer participation and demand side management are limited compared to west countries [11].

The development of the SG is specific to each country. As seen in this section, the development of a SG can only take place once the grid is robust. Making the grid strong and ready to keep up with the energy growth is the first step towards SG; this was the case in China and India. The countries can then focus on the optimization of the grid, by increasing its energy mix and including new IT and communication tools that can increase energy efficiency, as it is the case for the US. For the case of Turkey, the country still needs to improve its electrical grid, in order to reduce its transmission and distribution losses. Once the grid is secure, more improvement can be done in terms of optimization of the energy consumption of the country.

IV. SG IN MOROCCO

This section will describe the current situation of the Moroccan electric grid. It will also present the government initiatives that aim at facing the rising demand. Challenges and barriers on how Morocco can set its SG will be given at the end.

A. *Moroccan Current Electric Grid*

Morocco is the only North African country that does not currently possess oil or natural gas resources. The country needs to meet the rising energy demand which makes it dependent on foreign imports. The later puts the country in a vulnerable situation when it comes to energy security. In fact, the electric power consumption per capita went from 878 kWh in 2013 to 901 kWh in 2014 [19]. The electric power consumption per capita in the Middle East and North African countries reached 2,875 kWh in 2014, while it was 483 kWh in the Sub-Saharan and African countries [19]. The latter numbers show that Morocco is at an intermediate position compared to the two regions, which means that the country is making efforts to improve its energy sector. The total installed capacity in Morocco was 8,300 MW in 2016 of which 14% is based on renewables and 46.7% is based on coal [28]. In 2015, the total capacity was 8,158.5 MW with 13.4% from renewable and 49% from coal [29]. The projections show that Morocco is planning to reduce its coal based generation while increasing natural gas and renewable energy generation as depicted in Figure 2 [28][29][31].

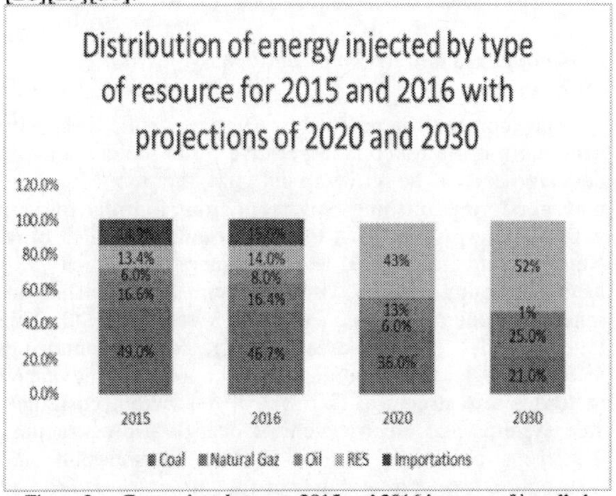

Figure 2. Comparison between 2015 and 2016 in terms of installed capacity resources with 2020 and 2030 projections

At the transmission level, Morocco contains 25,545 km of electric lines of very high and high voltage, 85,728 km of electric lines of medium voltage and 205,372 km of electric lines of low voltage [28]. The Moroccan electric grid is also connected to the Spanish and Algerian Electric grid via a 1,400 MW and 1,200 MW exchange capacity lines respectivly. The country is constently reinforcing its transmission sector to cope with the increased demand and the integration of renewables into its grid. In addition, Morocco is striving to becoming a hub that connects Europe to Africa.

At the distribution level, Morocco adopted SCADA as a management tool at the regional level. This can be considered as a first step toward a moroccan SG [30].

B. *Moroccan Government Initiatives*

Morocco took part in major initiatives that concern the protection of climate and environment, namely, he took part in the Rio de Janeiro, Johannesburg or Kyoto agreements. In 2009, the kingdom launched its national energy strategy, pointing out the importance of the power sector in the country.

The Moroccan High Commission for Planning (Haut Commissariat au Plan) published a report "Prospective Maroc 2030" [1], in which specific goals on the energy sector were set. The most important goal is reaching a total electric production coming from renewable resources of 52% by 2030 [1]. To achieve the latter, multiple projects started in different regions of the country, the plan is to add a total capacity of 6,760 MW between 2015 and 2025 (3,120 MW of solar, 2,740 MW of wind and 900 MW of hydro-power). Concerning solar energy, the first part of NOOR project in Ouarzazate is now operational with a total capacity of 160 MW. Like NOOR 1, NOOR 2 and 3 will use solar thermal technologies and will have a total capacity of 350 MW, while NOOR 4 will be based on photovoltaic technology with a total installed capacity of 70MW. For the wind energy, 800 MW is already operational, 550 MW under construction, and 850 MW is in the planning stage. In addition, the minister of energy stated that an additional 1,000 MW wind power is planned to be installed between 2012 and 2025 [13]. It is only natural that new laws and regulations arise to cope with the energy transition that the country is undergoing. Table 1 details the laws and regulation realted to RE that were issued since 2006 [14].

TABLE I. MOROCCAN LAWS

Law Name	Issue Date	Main aspects
Law 54-05	February 2006	Allows the government and the local authorities to delegate the manegement of electricity supply services to private entites
Law 13-09	June 2010	This law concerned renewable energies. It allowed generation by the private sector in medium and high voltage. The generation of electricity needed to fulffil some characteristics regarding the capacity of the generation of the farm; a declaration is necessary if the farm capacity

		is between 20kW and 2MW and an authorization is needed when the capacity is equal or higher than 2MW
Law 47-09	May 2012	Described energy efficiency as a fourth energy after fossil fuels, renewable energies and nuclear energy. The law also emphasized on the importance of the implementation of energy efficiency measures in all sectors
Law 57-09	June 2010	Concerned the creation of the Moroccan Agency for Solar Energy (MASEN) with the specific target to implement and reach the solar plan
Law 16-09	June 2010	Published for the creation of the National Agency for the Development of Renewable energies and Energy Efficiency (ADEREE). This institution was created in order to launch sectorial programs all over the country in order to seak the potential of each region in terms of renewable energies and energy efficiency so that new solar and wind projects can be installed in order to produce electricity
Law 58-15	December 2015	Amendment and completing law 13-09. This latter allows the generation of electricity at the low voltage. It also presents the possibility of selling excess of electrity, but the producer can not sell more than 20% of their total annual generation

Another Moroccan initiative is the program of rural electrification that started in 1996.

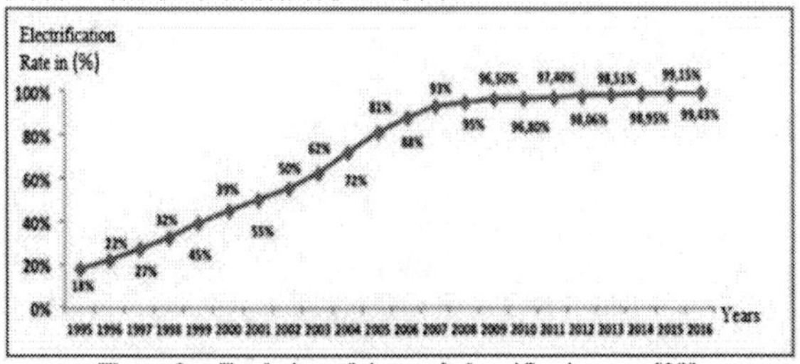

Figure 3. Evolution of the rural electrification rate [28]

At that time, the rural electrification rate was 18% only, at the end of 2015 this rate atteined 99.15%, Figure 3 presents how this rate kept increasing during these last years [28].

C. Moroccan Electricity Power Industry

Before the 1990's Morocco faced shortage in its electricity supply and part of the country experienced regular power cuts. It was then imperative to reform the electricity sector in order to mitigate the shortage issue. Independent Power Producers (IPPs) were authorized to privately generate electricity which was monopolized by ONEE before. The reform aimed at reducing the electricity generation using oil and hydropower and shifting some of it to coal and natural gas. The reform set by Morocco did not cover only generation but it was extended to distribution. The privatization of the distribution sector helped establish and improve the distribution infrastructure. At the transmission level, ONEE still has monopoly. As for generation ONEE contributed by 29.2% of the electricity generated in 2016, mainly from hydro, thermal and wind energy resources. IPPs, which are TAQA (also called JLEC (Jorf Lasfar Energy Company)), EET (Energie Electrique de Tahhadart), CED (Compagnie Eolienne du Détroit), TAREC (Tarfaya Energy Company) and MASEN contributed by a total of 56.9%. The remaining 14% of energy were imported from the interconnections between Spain and Algeria. At the distribution level, ONEE covers 45%, another 42% is covered by different public and private utilities. The remaining 12% are direct customers. Figure 4 details the situation of the Moroccan electrical industry in 2016 [28].

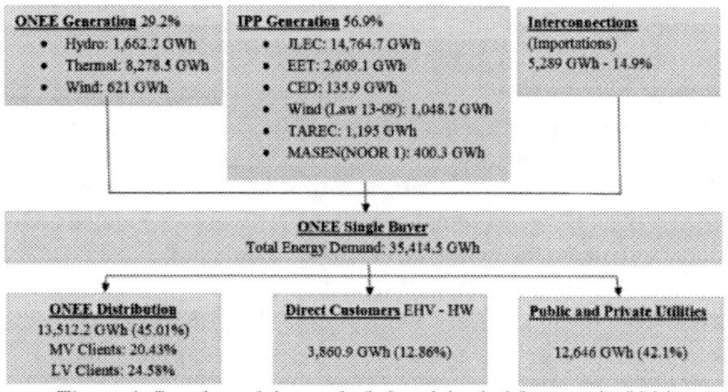

Figure 4. Supply and demand of electricity in Morocco in 2016

D. Challenges and Recommendations Toward a Moroccan SG

The inclusion of renewable energies in the electricity generation in Morocco is imperative to its socio-economic development. The government and the top Moroccan managers are publishing new regulations in order to cope with the energy transition that the country is undergoing. New projects are planned and the generation capacity will keep increasing to deal with the increasing demand, but renewable energies remain a small part of the SG. This latter needs a wide-area control and monitoring, information and communication technologies, advanced metering infrastructure, distributed generation, customer-side systems and electric vehicle charging infrastructure. The first challenge resides at the transmission and distribution levels; as a matter of fact, these losses reached a percentage of 14.7% in 2014, therefore they definitely need to be reduced to secure the grid [19]. The percentage has been increasing since 2004 where it was as low as 5% [19]. In order to reduce losses, a wide area monitoring network needs to be established. As far as the transmission infrastructure, more lines are being deployed each year to accompany the demand. At the distribution level, the metering infrastructure still relies on conventional (electro-magnetic) meters. It is clear that an AMI needs to be implemented. Given the 58-15 law, for instance, enabling low voltage integration of RE will require a metering infrastructure that will enable the utilities to bill the customers accordingly and to implement a particular pricing scheme. In addition, if Morocco expects to apply measures for peak shaving, it will be important to involve the customers which can only be achieved through an AMI. Introducing energy management systems and dynamic pricing could significantly improve energy efficiency and maximize the use of locally generated RE. The country will need to implement social awareness initiatives to promote the use of distributed generation. The importance of SG needs to be promoted as well and the different demand response programs need to be explained to the public in order to show the benefits customers can benefit from after installing an AMI. Security is also a challenge in implementing a SG. The country needs to procure the most secure network for its customers in order for them to trust the utilities and share their private data.

V. Comparison of Moroccan SG situation with other countries

In this section, we compare the current situation of Morocco to various countries using the main components of the SG as the metric for comparision. The main components of the SG are:

TABLE II. SMART GRID COMPARISON TABLE

		Morocco	**India**	**Turkey**	**US**	**China**
A	Electric Power Consumption per Capita (2014)	901.13 KWh per capita	805.60 KWh per capita	2,854.60 KWh per capita	12,986.74 KWh per capita	3,927.04 KWh per capita
	Regulations and perspectives	National energy strategy (2009)	SG roadmap for India (2013)	Strategic plan 2015-2019	The US SG initiative (2007)	Strategic plan for a robust grid (2009)
	Total installed capacity	8,300 MW (2016)	329,000 MW (2017)	80,000 MW (2016)	1,190,500 MW (2017)	1,645,750 MW (2017)
	Wide-Area Monitoring	SCADA	Phase Measurement Units	SCADA / GIS	Strong Area monitoring	Development of SG on HV and smart dispatch and control technologies
B	Transmission losses	14.7%	19%	15%	6%	6%
	Business Model of the electric industry	56.9% of generation from IPPs 42.1% of distribution is private	5% of distribution is private [32]	65% of generation from IPPs All distribution is privatized [33]	41% of generation from IPPs [34]	Generation dominated by 5 state-owned utilities and distribution by 3 state-owned operators [35]
C	% of energy from RE	14.0%	31.3%	35%	17.0%	24.8%
D	Smart meter	---	Development of smart meters	Automated Metering Reading	Automated Metering Infrastructure	Smart meter procurements ongoing [35]
E	Electrical vehicle	---	Shift to electrical vehicles expected by 2030 [21]	---	Development of plug-in hybrid technologies	Target of 1 Million electric cars by 2020 [27]
F	Customer-based technologies	---	Development of Demand Response programs	---	Development of customer based technologies	Demand Side Management programs

A- Wide Area Monitoring
B- Information and Communication Technology
C- Renewable and Distributed Generation Integration
D- Advanced Metering Infrastructure
E- Electrical Vehicle Charging Infrastructure
F- Customer-side Systems

As detailed in Table 2, Morocco needs to follow the footsteps of other countries that are on the right track to deploying a SG. It is clear from Table 2 that the US is the only country that is consedering all the components, China and India come after. With the integration of RE into the energy mix in Morocco, it is only natural that the country will transition to a grid that is smarter and more robust.

VI. Conclusion

SG is the future definition of electric grid around the world. Countries are making huge efforts in order to reach the necessary intelligence needed to optimize the use of its electricity. This paper presented the main components needed to acquire a robust electric grid, along with its challenges. Different countries around the world set short and long term plans that would help reach a certain intelligence needed for their electrical grid. The energy transition that Morocco is undergoing will only help hasten the development of a SG. The country is on the right track when it comes to passing the right laws that accompany the transition to a smart grid. Morocco is facing some real challenges to attain a certain level of intelligence, but following steps of developed countries can help getting its SG.

Acknowledgment

This paper's publication was supported by two projects. "Fes-Meknes Energy performance"; financed by the Region of Fes-Meknes and "Development of Smart Metering and an Energy Management System in Morocco"; financed by the National Center of Scientific and Technical Research (CNRST). The authors would like to thank both the Region and the Center for their help.

References

[1] High Commission for Planning, "Prospective Morocco 2030: Morocco's Energy Outlook Issues and Challenges," High Commission for Planning, Rabat, 2009.

[2] I. E. Agency, "Technology Roadmap: Smart Grids," International Energy Agency, Paris, 2011.

[3] S. Luthra, S. Kumar, R. Kharb, M. F. Ansari, and S. L. Shimmi, "Adoption of smart grid technologies: An analysis of interactions among barriers," *Elsevier,* vol. 33, no. Renewable and Sustainable Energy Reviews, pp. 554-565, 2014.

[4] M. F. Nejad, A. M. Saberian, H. Hizam, M. A. M. Radzi, and M. Z. A. Ab Kadir, "Application of Smart Power Grid in Developing Countries," in *IEEE 7th International Power Engineering and Optimization Conference*, Langkawi, Malaysia, 2013.

[5] I. Colak et al., "Smart grid opportunities and applications in Turkey," *Elsevier,* vol. 33, no. Renewable and Sustainable Energy Reviews, pp. 344-352, 2014.

[6] T. Atasoy et al., "Challenges & Opportunities towards Smart Grid in Turkey; Distribution System Operator Perspective," in *IEEE PES Innovative Smart Grid Technologies Europe (ISGT Europe)*, pp. 1-6, Istanbul, 2014.

[7] M. G. Simoes et al., "A Comparison of Smart Grid Technologies and Progresses in Europe and the U.S.," *IEEE Transactions on Industry Applications,* vol. 48, no. 4, pp. 1154-1162, July/August 2012.

[8] U. S. D. o. Energy, "2014 Smart Grid System Report: Report to Congress August 2014," United States Department of Energy, Washigton, DC, 2014.

[9] J. Lu, D. Xie, and Q. Ai, "Research on Smart Grid in China," in *IEEE T&D Asia*, pp. 1-4, 2009.

[10] Z. Ming, A. Sikaer, and H. Kelei, "The Future Development of Smart Grid in China," in *8th International Conference of Intelligent Computation Technology and Automation*, 2015.

[11] Z. Ruihua, D. Yumei, and L. Yuhong, "New Challenges to Power System Planning and Operation of Smart Grid Development in China," in *International Conference on Power System Technology*, pp. 1-8, 2010.

[12] Directorate of Observation and Programming, "Energy Statistics," Ministry of Energy, Mining, Water and Environment, Rabat, 2014.

[13] Ministry of Energy, Mining, Water and Environment, "Mr. Amara: "Morocco is developing energy projects with real investment opportunities valued at more than $ 37 billion"," 2015.

[14] Ministry of Energy, Mining, Water and Environment, "Texte Réglementaires: Electricité et Energies Renouvelables," Moroccan Kingdom.

[15] The Shift Project, "Data Portal: Browse Energy and Climate Data," The Shift Project, [Online]. Available: tsp-data-portal.org.

[16] Z. M. Fadlullah and N. Kato, Evolution of Smart Grids, New York: Springer International Publishing, 2015.

[17] I. L'hadi, "Liberalization of the Moroccan Electricity Market: A General Equilibrium Analysis," Ifrane, 2016.

[18] Indian Ministry of Power - Central Electricity Authority, "All Installed Capacity of India," 30 September 2017. [Online]. Available: http://cea.nic.in/reports/monthly/installedcapacity/2017/installed_capacity-09.pdf. [Retrieved April, 2018].

[19] World Bank Group, "The World Bank," IEA Statistics, 2014. [Online]. Available: http://data.worldbank.org/indicator/. [Retrieved April, 2018].

[20] Ministry of Power - Central Electricity Authority, "Annual Report 2014-15," Government of India, New Delhi, 2015.

[21] Ministry of Power , "Smart Grid Roadmap for India: Vision, Targets and Outcomes," Government of India, New Delhi, 2013.

[22] Republic of Turkey Ministry of Energy and Natural Resources, "Electricity," Ministry of Energy and Natural Resources, 2017. [Online]. Available: http://www.enerji.gov.tr/en-US/Pages/Electricity. [Retrieved April, 2018].

[23] Republic of Turkey Ministry of Energy and Natural Resources, "2015-2019 Strategic Plan," Ministry of Energy and Natural Resources, Ankara, 2015.

[24] P. Zummo, "America's Electricity Generation Capacity 2017 Update," American Public Power Association, February 2017.

[25] Z. XU, Y. Xue, and K. P. Wong, "Recent Advancements on Smart Grids in China," *Electric Power Components and Systems,* vol. 42, no. 3-4, pp. 251-261, 27 April 2014.

[26] Statista Inc., "Installed capacity of electric power generation in China in 2016, by source (in GW)," Statista, 2017. [Online]. Available: https://www.statista.com/statistics/302191/china-power-generation-installed-capacity-by-source/. [Retrieved April, 2018].

[27] " Energy Transition in the Power Sector in China: State of Affairs in 2016. Review on the Devlopments in 2016 and an Outlook. Analysis on behalf of Agora Energiewende and Chaine National Renewable Energy Centre," June 2017.

[28] ONEE, "Activity Report 2016," ONEE: Electricity Branch, 2017.

[29] ONEE, "Activity Report 2015," ONEE: Electricity Branch, 2016.

[30] S. Sahbani, H. Mahmoudi, A. Hasnaoui, and M. Kchikach, "Development prospect of smart grid in Morocco," *Procedia Computer Science,* vol. 83, pp. 1313-1320, 2016.

[31] Direction des Energies Renouvelables & de l'Efficacité Energétique, "Smart Grid: Un levier essentiel pour accompagner la montée en puissance des énergies renouvelables," Ministère de l'Energie, des Mines & du Développement Durable, Benguerir, 2017.

[32] International Trade Administration, "2016 Top Markets Report Smart Grid Country Case Study: India," U.S. Department of Commerce, 2016.

[33] International Trader Administration, "2016 Top Markets Report Smart Grid Country Case Study: Turkey," U.S. Department of Commerce, 2016.

[34] EIA, "Power Plant Operations Report," EIA, 2016.

[35] International Trade Administration, "2016 Top Markets Report Smart Grid Country Case Study: China," U.S. Department of Commerce, 2016.

Energy Management Policy with Demand Respond Method for a Smart Home

Win Thandar Soe, Cécile Belleudy
Department of Electronics
Laboratoire d'Electronique, Antennes et Télécommunications (LEAT), Université Côte d'Azur
Nice, France.
e-mail: win-thandar.soe@unice.fr, cecile.belleudy@unice.fr

***Abstract*—The increasing energy consumption increases the underlying environmental impacts. Hence, designing a more efficient energy management system is a vital research topic to address. A more innovative system for energy management makes energy consumption smarter and more efficient and reduces electricity bills. Several researchers proposed different home energy management systems with only grid supply as the source of energy. This study proposes a new online energy management approach that takes into consideration renewable energy sources coupled with battery autonomy. It is based on a novel architecture for collecting information on the energy consumption in a smart home via a wireless sensor network. In this approach, the first step is to identify the load type of each appliance based on its consumption. The second step consists in defining an energy management policy that utilizes the energy provided by the battery using a demand response method. These results show a reduction in the electricity bill for a smart home.

Keywords- demand response; energy management; load profile; smart plug network.

I. INTRODUCTION

In recent years, as the world's energy consumption has increased alarmingly, energy management systems have a critical role to contribute to the efficiency of modern smart homes. Many electrical devices used in homes do not optimize energy consumption, or to automatically or semi-automatically turn ON or OFF. From the electrical point of view, we can determine the total power consumption for a home, however, we cannot identify and observe the energy consumption of the different devices. To collect data from and manage the consumption individual appliances, smart plugs must be attached to each appliance. The plug then sends information to a gateway to be visualized on a dashboard. This collection data firstly identifies the appliance depending on its consumption. Using this classification, we will design an energy demand response system that reduces the required energy from the grid by using renewable energy to power one class of appliances.

The process of energy modeling and management in the smart home is based on three concepts:

- Energy harvesting [1][2]: local power resources, battery autonomy/charge, energy harvesting profiles according to the geographical climatic position,
- Home energy consumption [3][4]: classification and daily usage of appliances, lifestyle, load profiles during the day, time shifting, prediction and estimation of load analysis,
- Novel functionalities in Home Energy Management (HEM) approach [1][2][5]: integration of renewable energy sources and efficient battery management based on data collection and load profile classification, peak shaving energy management.

The aim of this article is to describe the load profile classification and the peak shaving energy management. This paper is organized as follows: First, Section II gives an overview of the related work. Section III describes a novel infrastructure for energy management system that classifies home appliances according to their power consumption. In Section IV, we propose and discuss an innovative energy management approach for a smart home that uses renewable energy sources to decrease the energy obtained from the grid. In Section V, we present the results of our preliminary experiments which are obtained by using our energy management approach. Finally, in Section VI, we conclude the paper and give some future work.

II. STATE OF THE ART

Electricity is the most inconstant and widely used form of energy. Therefore, electricity generation and consumption demand are growing globally. It is evident that among many other sources, the electricity generation is currently the largest single source of greenhouse gas emission, which is considered a significant contributor to climate change [6]. To mitigate the consequences of climate change, the current electrical system needs to undergo adjustments. The solution to these problems is not only in generating electricity more cleanly but also in optimizing the use of the available generating capacity. To achieve such optimization, the smart home could be an alternative. The intelligent control system inside a house introduces the concept of the smart home. Smart home technology provides an optimal alternative to its use for their security, comfort and energy saving as well. Previous studies have introduced different methods of home-based energy management system; however, most of these methods have considered grid supply. In the smart home environment, there are two primary methods to modify home energy consumption; total energy reduction with a price-based option and peak demand shedding or shifting method of Demand Response (DR) activities.

A. Problem Description

To fulfil the requirement of exceeding demands of energy consumption, different energy sources are installed in France. According to the study made in December 2016 [6] out of total capacity of 130 GW, approximately 48.3% installed energy is taken from nuclear sources, 35% is obtained from renewable energy sources and the fossil fuels and thermal energy sources contribute only 16.7%. Following the increased CO_2 emission in 2016, the nuclear power generation decreased due to the closure of several plants to conduct tests at the request of French nuclear safety authority. The energy generation from renewable sources covered nearly 20% of demand in France [6]. According to statistics presented by [6], the residential sectors are one of the main energy consumers which accounts approximately for 36% of the total French energy consumption. Seasonal variation and weather patterns reflect the energy consumption in France; for example, the use of energy is different during days of the week and time of day. France is the most temperature sensitive country in Europe. The demand for electric heating is higher in winter than in summer. Temperature below 15°C ultimately causes higher energy consumption. Similarly, the demand for user's activity is higher on weekdays than on weekends.

Energy consumption is one of the indices in determining the levels of development of a nation. The availability of energy supply to all sectors of life in any country causes the shortage of all kinds of energy, particularly electricity which is badly needed for economic development. Energy Management System (EMS) has been in existence in the energy sector for several decades. EMS also has age-long application as a vital role in the residential sector. Monitoring, controlling and optimizing the flow and use of energy are the key functions of energy management system [7]. HEM system also takes part in an essential role in performing residential demand response methods in the smart environment. It affords a homeowner the ability to automatically perform smart load controls based on utility programs, customer's preference and load priority.

B. Demand Response Approach

According to Federal Energy Regulatory Commission [8], demand response is defined as: "Change in electric use by demand-side resources from their normal consumption patterns in response to changes in the price of electricity, or to incentive payments designed to induce lower electricity use at times of high wholesale market prices or when system reliability is jeopardized". Time-of-Use (TOU) pricing is one of the price-based options under the classification of the U.S. Department of Energy [9].

- TOU pricing: a rate with different unit prices for usage during different blocks of time, usually defined for a 24-hour day. TOU pricing provides an incentive to shift loads from higher priced (On-peak) to lower-priced (Mid-peak) or (Off-peak) periods.

For example, Ontario Electricity TOU price periods [10] based on the schedule shown in Figure 1 follow: (i) Off-peak: 6.5 cents per kWh, (ii) Mid-peak: 9.5 cents per kWh, (iii) On-peak: 13.2 cents per kWh. TOU rates are typically reflected in the user's energy bill. In this study, we don't consider this price-based option.

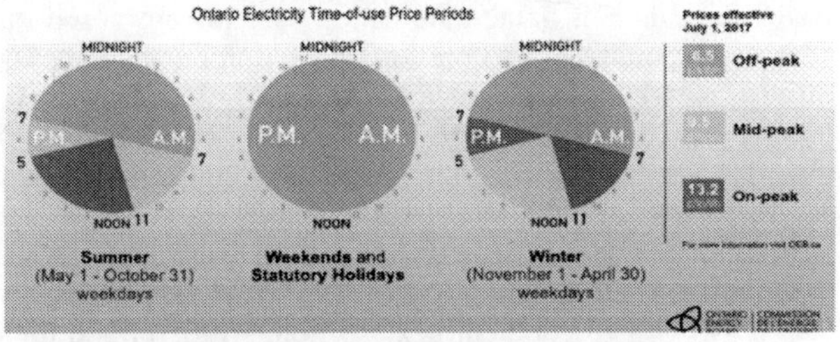

Figure 3. TOU Schedules and Rates **[10]**

Several HEM systems are based on different methods, such as load shifting, peak shaving etc.

- Load Shifting: a technique is used in demand-side management shown in Figure 2. It can be shifted the demand for a high power consumption of home appliances at different times within an hour or within a day when the price is lower.

Boynuegri et al. [1] have proposed HEM algorithm using load shifting method for the smart home. This algorithm includes the predefined criteria which combine grid ON/OFF situation, multi-rate electricity tariffs and the State of Charge (SoC) of batteries for all smart home platform with/without renewable energy sources.

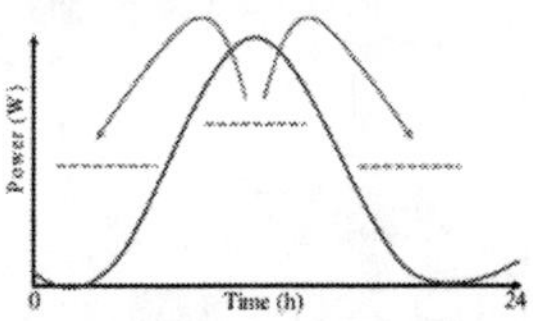

Figure 1. Load Shifting

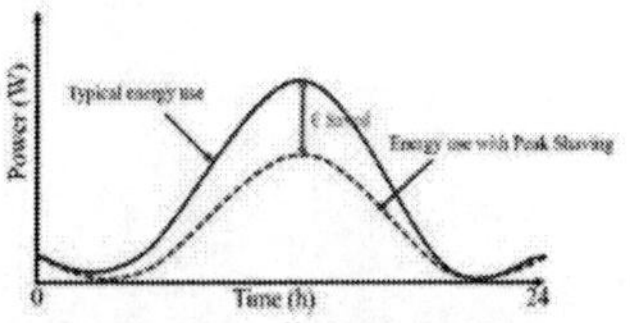

Figure 2. Peak Shaving Procedure

- Peak Shaving: a method which decreases total energy consumption, reduce the cost of electricity bill from the electricity grid due to the usage of renewable energy sources and shifting the higher power load to the night where the electric price is lower [5].

The procedure of peak shaving method is shown in Figure 3. Al-Saedi [5] presented the smart home peak shaving energy management system with a wired network that connects the house applications to a personal computer via USB port using K8055 interface board. Based on the peak shaving method in [5], the proposed policy with wireless network manages the household loads according to their preset priority and fixed the total household power consumption under a certain limit.

III. Energy Management Infrastructure and Load Classification

The home automation system has been used for the remote control of devices in the home since the 1990s and has formed the basis of the smart home building comfort [8]. A smart home system is divided into three main parts: a communication network, intelligent control and home automation [11]. The communication network can be created by wire, wireless or mixed communication technique between sensors and actuators. Many existing, well establishment home automation systems are based on wired communication. At present, the evolution and benefits of wireless technologies, such as Wi-Fi, wireless system are used every day and everywhere. The intelligent control intends to manage and monitor the entire house by internet services. Home automation that connects to the relevant smart components by using Home Area Network (HAN). Wireless sensor network system for smart home enables easy EMS construction. In next section, we described data collection and control architecture in the future smart home. Base on the principle of energy management infrastructure, we collected the data from home appliances by using the smart plug and classified the load.

A. Data Collection and Control Architecture in Smart Home

The smart home is a technology integration for greater comfort, autonomy, reduced cost, and energy saving as well. Figure 4 schematically illustrates the data collection and control architecture in the future smart home.

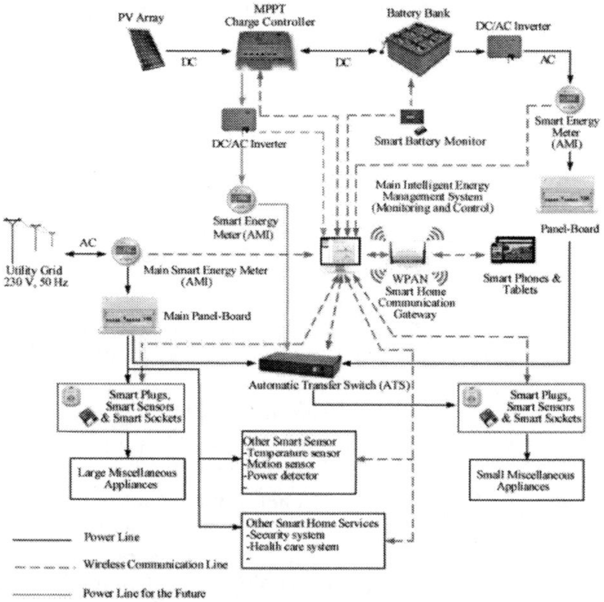

Figure 4. Data Collection and Control Architecture in Future Smart Home

In our proposed smart home, there are two supply sources for the household appliances: one source comes from the power grid and other from the renewable energy source, such as solar energy via battery. In this model, household appliances relate to an intelligent control system to communicate to the database and to manage the energy by Wireless Personal Area Network (WPAN), the smart sensors, the actuators and the smart plugs. The system can operate both in grid parallel mode and in stand-alone mode. All appliances are plugged into the wireless-enabled smart plugs. Thus, energy and power consumption for each appliance is monitored and stored in a distant location. Similarly, we can identify load profile of appliance by using the smart plug. Moreover, the smart plug can turn appliances ON/OFF remotely based on the commands from the central control. For example, the activation time of the appliances can be shifted if necessary.

To realize first measurements and validation of the proposed approach, we decide to use Fibaro smart plug [12]. The wireless communication protocol is Z-Wave technology. Fibaro smart plug is a bi-directional wireless system: current sensing and remote control. This plug has one function to check its status whether they are active or not. Each Fibaro network has its own unique network identification number (home ID), which allows identifying the different appliances and their activities. After Fibaro system is switched on, the location of its individual components is automatically updated in real-time through status confirmation signals received from devices operating in a mesh topology network. The unique feature of this smart plug is real-time energy consumption measuring through color changing, crystal LED ring.

B. Home Energy Management Database Structure

Information from energy harvesting (E_H), state of the battery charge (E_b), power and energy of each appliance (E_c) are collected and stored in the database. This collection is based on smart suitable sensors and actuators. Weather information, forecast or current, are derived from OpenWeatherMap service. The user activity is depending on the occupant's action plan and we propose to classify the underlying consumption according to load profile presented in next section. To achieve that, real-time consumption gathered by using smart plugs with proposed wireless communication technology. Energy management policy work on the database that contains real-time and prediction values by slot time. Slot time has been defined as part of a day according to the user activity. This slot time is to predict energy consumption and harvesting to propose a schedule of the usage of the battery according to energy sources status. The aim is to avoid energy harvesting loss, peak power, and prevent the outage. Figure 5 shows the principles of the proposed energy management infrastructure in a smart home.

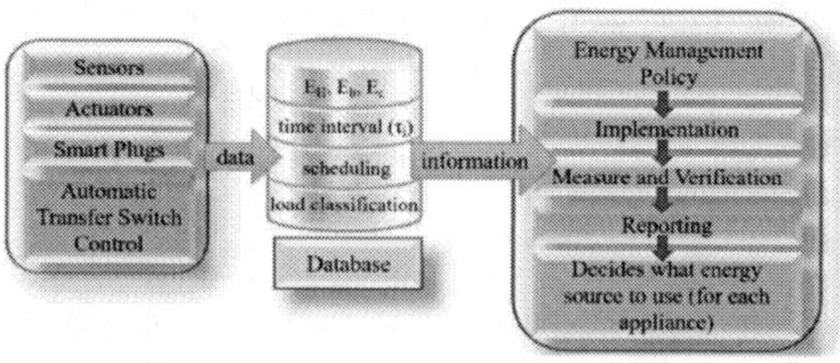

Figure 5. Principle of Energy Management Infrastructure

C. *Data Collection from Smart Plugs*

Fibaro smart plugs (Figure 6), which use a Z-Wave communication protocol, are in charge to collect power and energy consumption for the different home appliances. Real-time consumptions of appliances during each slot-time are collected and stored with Domoticz home automation application. This application allows visualizing and stored load profile for each appliance depending on the user's activity. On this base, our approach consists in determining a suitable classification of these load profiles with the aim to facilitate energy management.

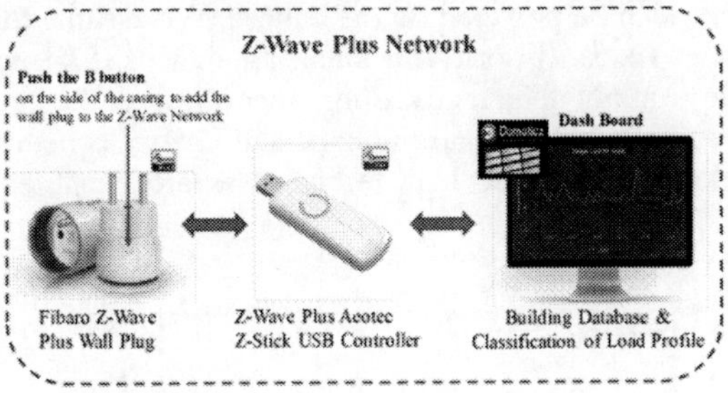

Figure 6. Installation to collect data from Fibaro Smart Plug

D. *Load Profiles and Classification*

The load profile is the variation in power consumption of an electrical load with time. The load profile is a specific concept in a smart home system that can vary depending, for example, on the user activity, season and weather condition etc. Appliances are drivers of residential power demands. Figure 7 demonstrates the aggregated load profiles of different appliances in a typical day. Our aim is to propose a suitable classification of these profiles to facilitate the energy management. As power consumption prediction is the main base for efficient energy management, the periodic load must be clearly identified and quantified. So, the proposed classification of loads is the following: intermittent load, phantom load, and continuous load [13].

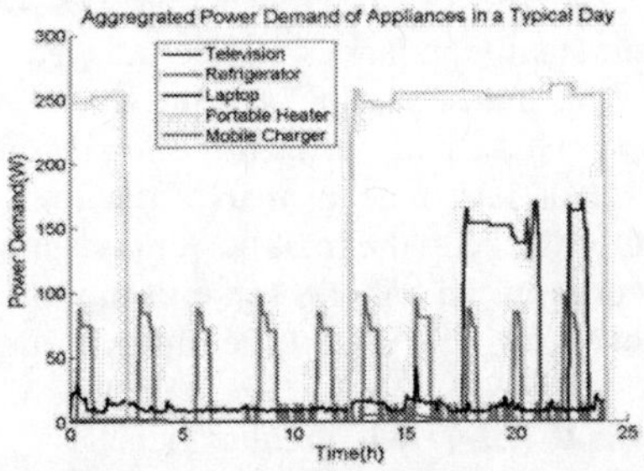

Figure 7. Aggregated Power Demand of Appliances in a Typical Day

1) Intermittent Load: The power consumption occurs occasionally or at a regular interval when an appliance is ON state. An appliance only operates a fraction of 24-hours periods depending on the user's activity. If the home appliances are active at a regular interval, these appliances' power consumption can be predictable but can present some variations which must be estimated with the confidence interval and degree. This load can be delayed eventually when the consumption of an appliance is lower than the total power consumption during peak hours or not enough energy.

2) Phantom Load: The phantom load or standby power also called wasted power occurs when the appliance is plugged into the socket and is not active (OFF/Idle states). In this state, the power consumption of home appliance is low almost zero. This function can define as a phantom load. After a while, this small amount of consumption tooks effect for the electricity bill.

3) Continuous Load: The certain appliances operate continuously or semi-periodically during a 24-hours period. The consumption can be variable according to appliance consumption mode. The characterization of each power mode must be done. This kind of load is highly predictable.

We defined a classification based on three kinds of load that can be combined to characterize the global power consumption. Each appliance is viewed as a combination of loads: one of the two main loads (continuous or intermittent load) and the phantom load which represents the small amount of power consumed in idle mode. Each load is defined by different parameters:

- Phantom Load: duration, the period with an interval of confidence, average maximum power. (Figure 8)
- Intermittent Load: duration with degree of confidence, the regularity with an interval of confidence, average maximum power. (Figure 9)
- Continuous Load: average duration, the period with an interval of confidence, average maximum power. (Figure 10)

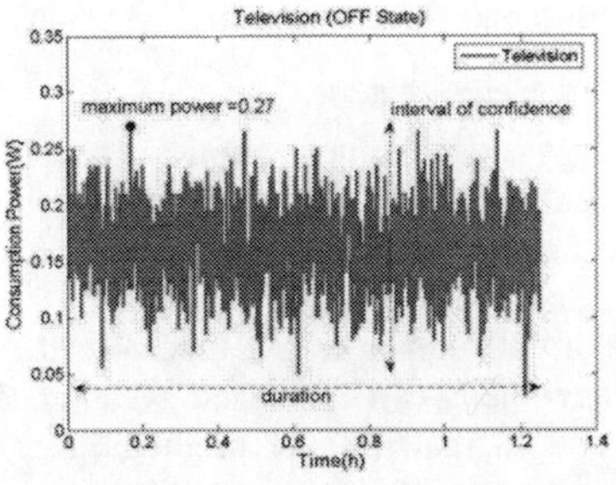

Figure 8. Measurements and Interpretation of Phantom Load

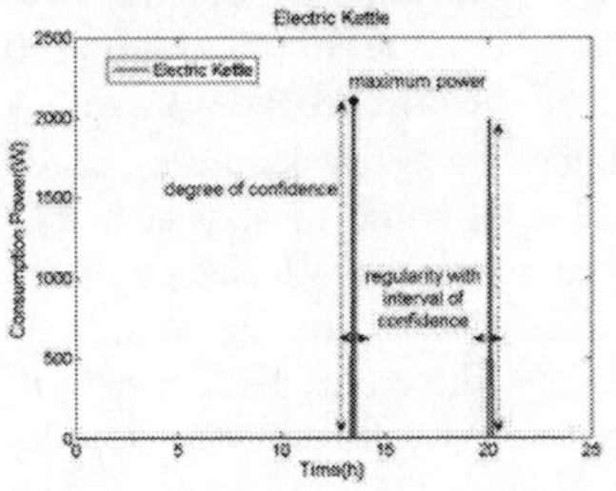

Figure 9. Measurement and Interpretation of Intermittent Load

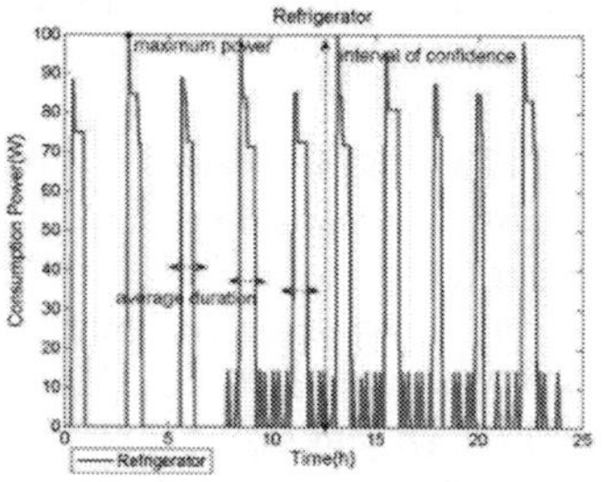

Figure 10. Measurement and Interpretation of Continuous Load

An automatic detection of the load has been implemented for real-time identification of data collected on the smart plugs. The first day for identification is needed. For instance, the home electricity usage of television at OFF state is represented in Figure 8. Where we analyzed the measurement and interpretation of the phantom load. Similarly, the intermittent load during a day as an example: Electric Kettle (ON state) is shown in Figure 9. The load profile of a refrigerator as an example of the continuous load is presented in Figure 10.

IV. ENERGY MANAGEMANT POLICY

In this paper, we propose an energy management policy using the peak-shaving method. The energy management policy is based on the different interval of time on which the schedule of battery charge, harvested energy and power consumption of appliances is built. On this basis, decision/action to reduce energy consumption from the grid are determined. The next subsections present the different parameters used and the proposed HEM policy concepts.

A. Analysis of Measurement

The energy management policy considers different sources of energy like grid and energy harvesting. To quantify the last quoted energy source, weather conditions are needed, and real-time and forecast data can get from web service or data center as like OpenWeatherMap service. Energy management can be based weather condition and occupants' action, real time or predictable. The associated parameters are:

- Weather condition (real-time and forecasted): Sunny hours per day and its irradiance, external temperatures (maximum, minimum),
- Kinds of the day: weekday/ weekend or holiday,
- Occupant's activity according to day: go to work or school at daytime or stay at home or not at home for the whole day based on the user's activity, prediction and motion detector.

By e-learning, different profile activity per day like wake up, small activity, lunch, non-activity, dinner, night, etc., are identified. This schedule is built at first on prediction and actualize on real time.

B. Proposed Home Energy Management Policy

In each time interval, the proposed HEM policy starts by gathering information, which includes the status (ON/OFF) and the power consumption of all appliances (watt, W), load priority (level) with its associated duration and the usable energy of the battery. A preliminary schedule is built based on the different prediction. When the decision is made, the HEM sends control signals to change the selected appliance status. In our study, we consider the three cases depending on the battery and electricity supply.

In the following, small appliances are defined by the fact that they can be powered by the battery. We assume that the customer fixes a priority for small appliances if he accepts to delay or not their activation. According to the level of energy in the battery and the expected energy consumption as described in Table I, different cases are feasible: (*b = battery, g = grid*)

TABLE I. SUMMARY OF PROPOSED HEM POLICY

Case	Conditions	b	g	Load Priority
1.	$E_{cs}(\tau) < [E_{b,\,usable}(\tau) + E_H(\tau)]$ SoC condition $\geq 70\%$	*		All small appliances are ON
2.	(a) $E_{cs}(\tau) > [E_{b,\,usable}(\tau) + E_H(\tau)]$ $50\% \leq$ SoC condition $\leq 70\%$	*		From the highest to lowest (with delay)
	(b) $E_{cs}(\tau) > [E_{b,\,usable}(\tau) + E_H(\tau)]$ $25\% \leq$ SoC condition $\leq 50\%$	*	*	From the highest to lowest
3.	$E_{cs}(\tau) > [E_{b,\,usable}(\tau) + E_H(\tau)]$ SoC condition $< 25\%$		*	All appliances are ON

Case 1: If the total energy consumption of small miscellaneous appliances $[E_{cs}(\tau)]$ is lower than the usable energy of the battery $[E_{b,\,usable}(\tau)]$ plus the total harvested energy $[E_H(\tau)]$ for and during a time interval, all small appliances are powered by the battery. The HEM will force the status of small miscellaneous appliances $[A_{s,i}$ (i= 1,2,3,…,n)] to be "ON". Assume that the state of the charge of the battery (SoC) is greater than equal to 70% condition. No action is taken from the grid.

Case 2 (a): If $[E_{cs}(\tau)]$ is greater than $[E_{b,\,usable}(\tau)]$ plus $[E_H(\tau)]$, some small appliances can be delayed or powered by the battery. In this state, SoC assumes between 70% and 50% condition. HEM will perform the lowest priority load to turn OFF and determine primarily the highest priority load to turn ON. To keep the total power consumption under the usable energy of the battery, some of the lowest priority load will delay for the next time interval until enough energy of the battery.

Case 2 (b): If $[E_{cs}(\tau)]$ is greater than $[E_{b,\,usable}(\tau)]$ plus $[E_H(\tau)]$ and SoC condition assumes between 50% and 25% condition, some small appliances will be powered by the battery and others by the electricity grid. In this case, the supply source for home appliances can be dual. With the performance of automatic transfer switch, some of the highest priority load will shift to use electricity supply due to the insufficient energy of the battery.

Case 3: If $[E_{cs}(\tau)]$ is greater than $[E_{b,\,usable}(\tau)]$ plus $[E_H(\tau)]$ at the SoC condition is below 25%, all appliances will be power from the grid to be more efficient energy and keep the usage of battery charge.

The proposed policy calculates the power of selected appliances and it will define state of each appliance either it is active or inactive. If the total power of these appliances exceeding the power limits, the use of this policy classifies the appliances and prioritize them in ascending order.

V. PRELIMINARY EXPERIMENTS

To process HEM, we recognize and identify different home appliances when coupled with smart plugs and monitor the progress at first day. We have collected data from appliances, such as quality of prediction and degree of confidence and classify different kinds of loads in our database. Table II describes the preliminary results of selected home appliances. We discussed the different parameters of appliances and defined in section III. The experimental time for each appliance was 24 hours. Quality of prediction is denoted by numbers from 0 to 5 indicating from lower (0) to high (1) priority. *(C = Continuous Load, P = Phantom Load, I = Intermittent Load)*

TABLE II. RESULTS FROM PRELIMINARY EXPERIMENTS OF DIFFERENT KIND OF LOADS

Plug No.	Appliances	C	P	I (Regular)	I (Irregular)	Quality of Prediction
1.	Refrigerator	*	*			5
2.	Light		*	*		4
3.	PC, Laptop		*		*	3
4.	Multimedia		*		*	2
5.	Coffee Maker		*	*		1
6.	Charger		*		*	0

These preliminary tests have been made to support home appliances for standard size family and this proposed HEM system allows approximately 28% reduction in total grid energy consumption. These results could be extended to larger HEM systems to provide and prove the energy efficiency of this proposed approach.

VI. CONCLUSION AND FUTURE WORK

To measure the energy consumption in the smart home and analyze their features, we propose a data collection architecture base on smart plugs and the definition of three types of load. The aim is to improve the prediction of the energy consumption which is estimated at intervals of time that compose a day. According to these different load profiles, we propose to modify the functionality of the energy management policy for a smart home to integrate with renewable energy sources with the aim to optimize the use of the battery. Although much demand response methods are widely implemented by commercial and industrial side, nowadays it is enabled in the residential sector. Depend on the user activity and current electricity policy, to integrate peak shaving method with the smart appliances for the smart home. The goal of our research is to solve outage of electricity from the grid, to seek energy from the local power resources and reducing of electricity bill by using per month.

Base on the above conditions, the next step of work is to validate our approach with wider experiments. The second aim is to couple the energy management policy and a simulation framework which will represent the sensor network in charge to identify the load activity. The final aim is to provide a tool to realize a conjoint architecture exploration on suitable wireless sensor network and energy management policy in the smart home.

ACKNOWLEDGMENT

The authors would like to acknowledge by the mobility funded by Erasmus Mundus Mobility with Asia (EMMA) in the framework of EU Erasmus Mundus (Action 2). We also express our gratitude to Laboratories d'Electronique, Antennes et Télécommunications (LEAT).

REFERENCES

[1] A. R. Boynuegri, B. Yagcitekin, M. Baysal, A. Karakas, and M. Uzunoglu, "Energy management algorithm for smart home with renewable energy sources," 2013, pp. 1753–1758.

[2] A. Saha *et al.*, "A Home Energy Management Algorithm in a Smart House Integrated with Renewable Energy," p. 6.

[3] T. Zhu, A. Mishra, D. Irwin, N. Sharma, P. Shenoy, and D. Towsley, "The case for efficient renewable energy management in smart homes," in *Proceedings of the Third ACM Workshop on Embedded Sensing Systems for Energy-Efficiency in Buildings*, 2011, pp. 67–72.

[4] M. Pipattanasomporn, M. Kuzlu, and S. Rahman, "An Algorithm for Intelligent Home Energy Management and Demand Response Analysis," *IEEE Trans. Smart Grid*, vol. 3, no. 4, pp. 2166–2173, Dec. 2012.

[5] F. A. T. Al-Saedi, "Peak shaving energy management system for smart house," *Int. J. Sci. Eng. Comput. Technol.*, vol. 3, no. 10, pp. 359–366, 2013.

[6] "2016 Annual Electricity Report: Bilan électrique 2016." [Online]. Available: http://bilan-electrique-2016.rte-france.com/mon-bilan-electrique-2016-en/. [Accessed: 05-Mar-2018].

[7] M. Amer, A. M. El-Zonkoly, A. Naamane, and N. K. M'Sirdi, "Smart Home Energy Management System for Peak Average Ratio Reduction," *Ann. Univ. Craiova Electr. Eng. Ser.*, vol. 38, pp. 180–188, 2014.

[8] D. Wight *et al.*, "2010 Assessment of Demand Response and Advanced Metering - Staff Report," *Fed. Energy Regul. Comm.*, vol. 2, pp. 1–117, 2011.

[9] Q. Qdr, "Benefits of demand response in electricity markets and recommendations for achieving them," *US Dept Energy Wash. DC USA Tech Rep*, no. 2, pp. 1–122, 2006.

[10] "Electricity rates | Ontario Energy Board." [Online]. Available: https://www.oeb.ca/rates-and-your-bill/electricity-rates. [Accessed: 05-Mar-2018].

[11] O. Elma and U. S. Selamoğullari, "A home energy management algorithm with smart plug for maximized customer comfort," in *Electric Power and Energy Conversion Systems (EPECS), 2015 4th International Conference on*, 2015, pp. 1–4.

[12] "Wall Plug | FIBARO Manuals." [Online]. Available: https://manuals.fibaro.com/wall-plug/. [Accessed: 13-Oct-2017].

[13] W. T. Soe, I. Mpawenimana, M. Di Fazio, C. Belleudy, and A. Z. Ya, "Energy Management System and Interactive Functions of Smart Plug for Smart Home," *Int. J. Electr. Comput. Energ. Electron. Commun. Eng.*, vol. 11, no. 7, pp. 884–891, 2017.

Energy Planning Support with Geomapping Tool and Energy Demand Estimation: The Energis Platform

Gema Hernández Moral, Víctor Iván Serna, Giulia Massa and César Valmaseda
Energy Department
Fundación CARTIF
Boecillo, Valladolid, Spain
Email - gemher@cartif.es, vicser@cartif.es, giumas@cartif.es, cesval@cartif.es

***Abstract*—Energy directives aim at decreasing energy consumption and assuring a low carbon environment, in line with climate change mitigation strategies. Their implementation gives rise to the need of energy plans to improve actual energy tendencies. However, making informed decisions in energy planning is not always immediate, nor it is based on contrasted criteria, since normally the current conditions are unknown. There is a lack of tools that can aid planners in this decision-making process by showing the energy status of a determined area. In order to provide an adequate diagnosis and propose energy actions to cope with those needs, the baseline conditions should be based on validated calculation methodologies that make use of accurate and reliable data and are displayed in an easy and understandable manner. To deal with these problems, the ENERGIS platform is proposed where the energy demand of the residential sector is estimated with validated Energy Performance Certificate (EPC) software tool in Spain making use of public available data of buildings from official sources. The results are then reflected on a map, making use of geo-visualisation capabilities, which sets the basis for informed energy planning.**

Keywords- Energy planning; demand estimation; Energy Performance Certificates (EPCs); maps; geo-visualisation; Geographic Information System (GIS).

I. INTRODUCTION: ENERGY PLANNING CONTEXT IN EUROPE AND PLATFORM NEED

Growing CO_2 emissions have increased the concerns with respect to Climate Change, which has been especially acknowledged by the United Nations with the signature of the Paris agreement [1]. In this ambitious, universal agreement some guidelines and objectives were set to provide a way forward to fight against climate change and to stop the current temperature rise below 2 degrees.

The main source of these emissions is the deployment of fossil fuels for heating and cooling purposes, which in the end can be translated to the energy consumption produced by this type of fuels. In Europe, one of the main sectors contributing to the increase of CO_2 emissions is the built environment, and in particular the residential sector accounting for 25,4% of the share according to Eurostat figures.

These facts describe a reality where it is necessary to act upon, to implement regulations that enable to control the energy consumption and thus be able to secure a low carbon environment, which in the end will contribute to the fight against climate change.

Reacting upon this necessity, the European Commission has proposed a package of Energy Directives [2], which aims for three main goals: putting energy efficiency first, achieving global leadership in renewable energies and providing a fair deal for consumers. It includes as well eight different legislative proposals that tackle, among other Energy Efficiency, Energy Performance in Buildings, Renewable Energy and Governance.

These directives set certain objectives to Member States and should be transposed by each nation in order to comply with them by establishing plans and strategies. Depending on the administrative structure existing in each country, the plans can either be established at national level, or some high level guidelines at national level can be set and then specific objectives or actions at regional level implemented. After each Member State has carried out his strategies, the results should be reported at European level.

These steps in energy policy implementation require the establishment of mechanisms to be able to propose adequate strategies according to the current energy status of the residential sector in each country at regional and national level. It is paramount to base the energy actions on a precise diagnostic of the situation to avoid off-target actions that result in a non-compliance of the objectives set and an inadequate allocation of resources.

However, when a national or a regional authority intends to perform any kind of energy planning there is a lack of information and tools related to energy that aid in the diagnosis of the current situation at a global level (district, city or regional level), which in the end hinders the possibility to propose any contrasted energy action plan.

Nevertheless, there are many elements that are not fully exploited and that can be deployed in this field. Firstly, there is public information regarding the built environment and climatic conditions, such as the cadastre or climate data from weather stations. Secondly, from the need imposed at European level to submit EPCs, several validated methodologies or tools have emerged at national level in order to obtain Energy Labels and basic energy information.

Thirdly, the existing Geographic Information Systems (GIS) can provide the spatial component, which is intrinsic to any kind of planning that involves considering cities, regions or nations.

Based on these existing elements and combining their potential, the ENERGIS platform offers support in the energy planning process by enabling the visualization through GIS of the energy demand of the residential sector, which has been calculated deploying validated methodologies (Energy Performance Certificate tools in Spain) using public information. This platform will set the basis to have an adequate view of the current status and be able to make informed decisions.

In Section II a review of similar approaches to ENERGIS has been performed. Then, in Section III, the approach followed in the project is presented. The specific modules integrating the platform are described in Sections IV, V and VI. Finally, some conclusions are extracted in Section VII.

II. SOME APPROACHES IN THE ENERGY PLANNING FIELD

The need to represent the energy current status has been explored in some studies in this field and they should be highlighted since some of the main elements introduced in the ENERGIS platform are addressed, even though lacking some of the features of the latter. These major differences will be explained in Section II-B.

A. *Examples of energy mapping tools*

The following represent examples that integrate simplified estimations to obtain energy indicators and are combined with GIS, which enables an immediate visualization, as it will be observed in the examples.

1) Estimated total annual building energy consumption at the Block and Lot level for NYC

Developed by the Sustainable Engineering Lab, this project aims at analysing the dynamics of final energy consumption in the city of New York [7]. To this aim, the final heating, cooling, DHW and electricity energy consumption in the built environment is estimated. The results are then displayed following a colour code through a web map, providing two levels of aggregation: at block and at lot level. The information is complemented with a pop-up that appears when clicking on a block or lot.

The calculation methodology deployed was based on data at district level on the energy use, natural gas, diesel and vapour consumption of 2009 and it was combined with information coming from MapPLUTO, a geographical database of the Urban Planning Department in NYC. The estimation was based on the functions contained in the buildings (residential 1-4 people, residential multi-family, educational, health, storage, offices, commercial), making some special considerations regarding the location of the building. It is also based on the hypothesis that all ground floors have a commercial use and the upper floors are dedicated either to offices or residential use.

With respect to visualization, it was developed with Mapbox tool, where with a colour code the different energy intensities are represented. However, this map only covers the city of New York, where the calculations have been estimated, without offering the capability to replicate this methodology in other cities.

2) Energie label atlas

The objective of the Energie Label Atlas is to represent the estimated Energy Labels of buildings in Holland [9]. This project was carried out with the aim of covering almost all residential buildings in the country and of offering citizens the possibility to obtain an accurate EPC of their dwelling. This would require performing additional calculations over the estimated value proposed in the map, but at a lower cost than usual.

The origins of this project are to be found in the 'Block by Block – Brooklyn's Past and Present' project that showed this area in New York with a colour code representing the age of the building. This idea was replicated in the Netherlands within the project Smart CitySDK, where all the buildings in the Netherlands are assigned a colour according to their age [8]. Building upon it, the Energie Label Atlas emerged, assigning to each building or dwelling an estimated Energy Label.

To make these estimations reference buildings were calculated and a label was assigned to each of them. That implied the study of a number of building typologies that was representative enough of the residential sector in the Netherlands. Afterwards, all of the reference buildings were simulated, assigned to their corresponding typologies and mapped. In addition, the real Energy Labels coming from real EPCs were also mapped and the comparison among both results enabled. This results in some cases in high discrepancies between both results.

The visualization capabilities are similar to those of the NYC case (Section II-A-1); however, the whole country was covered providing the estimated results at block level, regardless of the visualization scale. The information is also complemented by a widget, which serves to detail information at dwelling level, if real EPC information is available.

B. *Main differences with the ENERGIS tool*

The projects described above represent interesting approaches to the energy planning problem, by providing basic information on the energy status of a city (in the case of NYC) or of a country (in the case of the Netherlands). However, it is worth highlighting the main differences to the ENERGIS platform.

The **calculation methodology** is considered highly relevant in order to provide accurate results. Therefore, instead of deploying estimated approaches in order to calculate consumption (as in the NYC case) the ENERGIS project is constrained to the **estimation of the energy demand**. This is due to the fact that there is no source of public information that can provide data on the energy systems, rendering it inaccurate to assume the existence of a determined energy system. The approach used in NYC to calculate consumption, that is, assigning an energy system according to the typology of the building would be highly beneficial if applied within ENERGIS and would improve the quality of the platform. However, since there is no available database where these systems are described in Spain, no reliability can be assumed from this process. Moreover, the calculation is focused on the **automation of validated tools** to generate EPCs in Spain. This fact ensures a determined level of precision to the results obtained, opposed to the abovementioned approaches. With respect to the **scale** considered, aggregation possibilities are granted as in the case of the NYC map, but the unit of measure used to estimate the demand is the block, as in both showcased projects. Nevertheless, the scope to which the ENERGIS platform can be applied is the same as in the Netherlands case: at country level, since the main source of data is the Spanish Cadastre which covers the whole country [10].

C. *The approach followed in the ENERGIS project – Main objectives*

As it has been observed, the ENERGIS key objectives can be summarized in three main points: (1) exploiting the potential of available public data sources, (2) automatizing a process of demand estimation based on validated methodologies, thus granting trust and precision and (3) exploiting visualization capabilities making use of GIS in the web.

In order to fulfil these objectives, the ENERGIS platform is divided into three main modules:

- Module 1: Information processing and treatment: which will be in charge of collecting the required information and processing it for its later use by Module 2.
- Module 2: Estimation of the energy demand: based on Energy Performance Certificates tools in Spain, this module will be in charge of automatizing the calculations and providing the results of the energy demand of the buildings at block level.
- Module 3: Geo-referencing and visualization: the results obtained by Module 2 are then displayed in a web map, following a colour code that corresponds to the Energy Label scale and providing different data to be shown and a range of filtering capabilities.

III. MODULE 1: INFORMATION PROCESSING AND TREATMENT

In order to design and develop this module, two main processes were carried out: identifying the main data sources to fulfil the needs of the second module (estimation of the energy demand) and the methods to process and transform the data.

A. *Identification of main sources of data*

Public data sources were examined in order to determine the most valuable among them according to the identified needs of Module 2. Information regarding the building is required (geometrical aspects, orientation, neighbouring blocks, thermal properties of the elements it is composed of and identification data) as well as climate data according to its location.

For this purpose the Spanish Cadastre was analysed. It contains relevant data on buildings which is accessible for download through standardized mechanisms, following directives on spatial data, as it is the INSPIRE Directive. From this data source, relevant information on geometrical aspects, orientation, identification data and neighbouring blocks can be derived.

To cope with the climate requirements, the National Code for Building Construction in Spain was queried, since it establishes reference climate data according to the province or town (static).

For the building thermal properties assignation the National Building Code was also consulted. Based on several studies a Catalogue of building elements and materials has been generated. Based on it the different EPC tools are able to assign thermal property values to walls (external and internal), roofs, floors, windows, etc. The ENERGIS project based different building characteristics on the values used in the EPC tools, where according to the type of element, the year of construction and the climatic zone, some thermal characteristics and other parameter were assigned.

B. *Processing of data*

The processing of the data was needed mainly to derive geometrical characteristics of the data contained in the cadastre. For instance, the calculation of the number of pillars to calculate thermal bridges and estimate the number of windows; shadow calculation based on neighbouring building blocks, to determine the orientation of a wall or if it is interior or exterior, etc.

All of these calculations were performed by deploying two scripts developed by VOXEL3D, one devoted to the extraction and pre-processing of raw data coming from the cadastre and the other devoted to the detection of neighbouring buildings and generation of shadow patterns that will be assigned to each individual wall and opening.

As a result of the processing of this data a JSON file is generated containing all the required categories of information, which is then sent to Module 2 to estimate the energy demand.

IV. MODULE 2: ESTIMATION OF THE ENERGY DEMAND

The second module can be considered the core of the platform. If inadequate calculation methodologies are deployed, then the complete platform loses veracity and thus cannot be deployed with the expected results. For visualization purposes, it should also be able to perform quick estimations. Therefore, the steps followed in the design and development of this module were: the definition of the requirements, definition of the module, its development and its validation.

A. Status of EPC tools in Spain

Having an understanding of the validated calculation methodologies deployed in Spain is paramount prior to the selection of one of them. Therefore, the first step was to perform an analysis of the tools used at national level.

Based on the requirements imposed by the Energy Performance of Buildings Directive (EPBD, 2010/31/EU) [3], Member States of the EU are required to make mandatory the submission of EPCs for every dwelling, building block, or commercial premise to be leased or sold, as well as for every new construction and public buildings.

In order to assure coherence among the results obtained in each Member State a methodological framework is described in the annex of the aforementioned Directive, which does not exactly set the formulas to be deployed, but instead presents the type of calculations to perform or which aspects to consider (for instance, thermal bridges).

Afterwards, each Member State has the obligation to transpose this framework in their country and develop either a concrete methodology or develop specific tools to serve this purpose. In the case of Spain four validated tools were developed and made available to the public free of charge: *Herramienta Unificada Líder-Calener* (HULC), CE3, CE3X and CERMA. The main differences among them are the following:

- **HULC**: tool developed based on public initiative, it presents the general energy certification method for buildings in the design phase, built buildings and existing ones (single family houses, building blocks, individual dwellings in a block and tertiary buildings). Through a relatively complex graphical user interface, the data insertion through this platform requires modelling in 3D and inserting precise data of the components integrating an element in the building (for instance the layers in a wall). The order followed when inserting the data should not be modified and the evaluation of different energy conservation measures proposed as an improvement to the baseline scenario cannot be evaluated.
- **CE3**: tool developed based on public initiative, presents a simplified method for the energy certification for existing buildings (single family houses, building blocks, individual dwellings in a block and tertiary buildings). The insertion of data at different levels of detail is allowed and if some data is unknown a library of reference data is provided. In addition, it automatically offers three energy conservation measures as a suggestion for its improvement.
- **CE3X**: similarly to CE3, this tool was developed based on public initiative, presents a simplified method for the energy certification for existing buildings (single family houses, building blocks, individual dwellings in a block and tertiary buildings). Through user-friendly graphical user interfaces the insertion of different levels of detail is allowed and also the evaluation of energy conservation measures. It offers interoperability capabilities with more advanced programs and the possibility to export an XML file with the results of the calculations.
- **CERMA**: the only recognized tool developed based on private initiative, which offers a simplified energy certification method for building blocks. It allows certifying single family houses, building blocks or individual dwellings in a block, for existing and new buildings. The evaluation of different energy conservation measures is allowed. Data requirements of this tool are high in comparison to the rest.

B. Definition of requirements

Based on the objectives defined for the platform, the requirements of the estimation of the demand module were quite straightforward: the tool should enable the introduction of an adequate amount of data that can be fed by the public data sources identified.

For this aim the study of the tools was performed, choosing in the end CE3X because of being the reference in terms of data model at national level (the data model proposed at national level to represent information related to EPCs follows the same structure as the output of CE3X). Moreover, it is easy to use and has data requirements that can be easily covered with information coming from public sources. Additionally, the accuracy behind the results provided in the calculations of the tools was also tested, which is explained in the validations named "prior to the selection of the EPC calculation tool" shown in Section E-1.

C. Definition and design of the module

The definition of the module was performed following the next steps:

- **Identification of the data required to perform simplified calculations**: based on the analysis performed on the data required by the estimation tool and the available data in public data sources, as well as the performance validations carried out, the exact datasets required were identified. Among these, data from the cadastre, three climatic zones maps and data on building elements contained in a catalogue were identified (walls, roofs, windows, floors, internal walls, thermal bridges, etc.).The data introduced into the platform is considered to be slowly evolving or almost static. For instance, in the case of the climate data, reference data is assumed at this stage of the project. Given a determined location, and the zone where it is located, certain climatic characteristics are deployed, which are tabulated according to the Building Code in Spain. Other data deployed in the platform, such as the geometrical data, is considered to be invariable; however, the datasets contained in the platform should be updated regularly according to the updates followed in the main source of data deployed: the Spanish Cadastre (approximately six months).

- **Simplified approaches for geometric definition of data**: from the data coming from the cadastre some could be directly applicable, while other geometric characteristics had to be estimated, since they were not directly available in the cadastre. For instance, the information on openings, thermal bridges or the shadow calculation.
- **Establishing a workflow diagram**: once the method to process the data had been established, a workflow depicting the functioning of the module was devised. In it the normal input data process and certification steps of the tool were taken into account.

D. *Development*

The development of the module was divided into two sub-modules. The first one is in charge of extracting the information contained in the JSON file coming from module 1 and generating one file per building block compliant with CE3X (.cex). Afterwards, the second sub-module will gather all of the compliant files and will automatize the CE3X process by introducing the information contained in the files and generating the Energy Performance Certificates. As a result, an XML is derived per building block where the heating, cooling and global demands are contained.

E. *Validation of the module*

Several types of validations have been performed during the design, development and validation process of the ENERGIS platform. Firstly, the appropriate EPC tool to use as a basis for the calculation module was chosen according to their data insertion processes, the capabilities they offer, and the accuracy obtained. Secondly, once the tool had been selected, hypothesis and simplifications with respect to geometry issues such as thermal bridges, windows, etc. have been tested. These hypotheses were based on a set of tests performed on a model by model comparison, where the objective was to test how much do the definition of thermal bridges, windows and other elements of the envelope affect the results. Finally, once the platform had been developed, the final results obtained with it when automatically retrieving the data from the public data sources, processing it and automatizing the process were analysed.

1) Prior to the selection of the EPC calculation tool

The objective of these validations was to test the functioning of the different tools and the results obtained among each other. From this analysis the main output was that even though the process followed in the data insertion process and the tools were different, the results were similar among each other. This allowed selecting the tool based on other type of criteria than only on accuracy aspects. The outcome of this process was the selection of CE3X as the main calculation tool to be deployed, as previously justified.

2) Testing the capabilities and how the program works

After having selected the calculation tool and having analyzed both the data required and the available data in public data sources it became apparent that some of the geometric data was not directly available in public data sources. Therefore, an estimation to calculate openings, thermal bridges and shadows was required.

To test if these hypotheses were reasonable in terms of precision deviation (i.e. their Energy Label did not vary) from a real Energy Performance Certificate, real cases were tested. These real cases were selected buildings in Valladolid (Spain), of which detailed data to perform energy simulations, was available. The EPC of real buildings was calculated based on information coming from detailed CAD plans, as if they had been performed by a certifier. The numbers of elements, their surfaces, as well as the final EPC results were contrasted and the result was satisfactory in terms of accuracy. As a consequence, and without delving into the hypotheses were calibrated and developed into the platform.

3) Testing the results obtained

Finally, once the hypothesis had been validated from an accuracy point of view, it was necessary to test that the process designed was responding as expected. To this end, the real calculations previously performed were compared to the results that would have been obtained manually by performing the geometric estimations proposed and also with the results obtained from Module 1. Several issues were detected during the process and errors were solved in a wide range of cases were different typologies of buildings and configurations were considered.

V. MODULE 3: GEO-REFERENCING AND VISUALISATION

The last module is devoted to the geo-referencing and visualization of data, which is highly relevant since it is the point where the information is communicated to the planner. A web-based platform has been developed taking into account the following aspects: the data to be stored and shown, how to store these data and the GIS web viewer.

A. *Visualization data*

The information that has to be shown in the visualizer and stored in the database is:

- Information of the buildings from the cadastre without processing: year of construction, address, use of the building, location, etc.
- Geometrical information of the building post-processed from the original geographical information: surface of the building, information about the façades and other elements, etc.
- Energy demand information calculated in the module 2: cooling, heating and global demand.

In order to be able to represent the data in a map using a GIS system it was necessary to populate a geodatabase which stores the aforementioned data with their location information.

B. *ENERGIS web mapping platform*

The ENERGIS web mapping platform is a map service platform that offers the geospatial information stored in the geodatabase by a set advanced tools for the access and management of this information, which is supported by vector cartography of the zones of interest or by orto-photographs coming from the *Plan National de Ortofotografía Aérea* (PNOA).

The backbone of the geographical web viewer is its functioning as an API that allows for flexibility, and easiness in the integration. This API allows consuming directly vector layers with energy information of urban settings. The capabilities offered can be summarized in the following: precise visualization tools at different scales; intelligent real-time filtering tools of the visualized contents, interactive querying the contents relative to each element of the map, showing different type of data, visualization with different base maps: PNOA, Open Street Maps, Mapzen or Stamen; and integration of other data of interest, such as the cadastre information or other post-processed information.

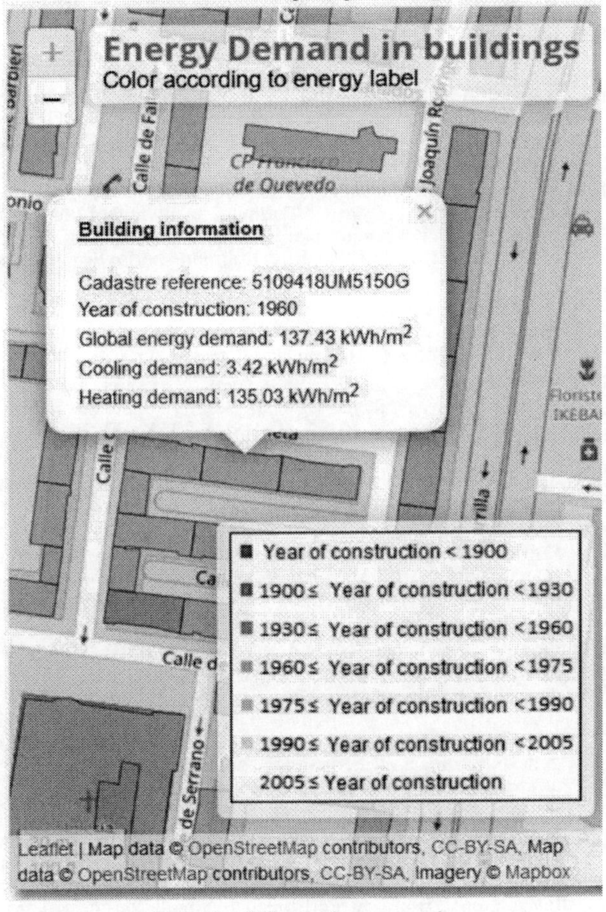

Figure 1. GIS web-viewer example

The values displayed through the viewer are represented through a colour code that corresponds to the EPC scale. This enables a quick understanding of the status of a city, identifying which areas are most in need of retrofitting. An example of the viewer can be seen in Figure 1. In it, some of the capabilities listed above can be observed.

As it can be seen, complementary information of each building can be displayed on the screen, such as: cadastre reference, year of construction, global energy demand, cooling demand and heating demand. In addition to this, more in depth information related to each building is available for download by the user. In it, data on the envelope used for its calculations, as well as additional identification data, thermal data and shadow information. Additionally, it should be mentioned that depending on the scale there are aggregation mechanisms where values of energy demand at region, district or at group of buildings' level can be displayed.

VI. LESSONS LEARNED

The development of an energy platform to support energy planning has offered a number of lessons learned, which have arisen from trying to implement the objective of working with public data as a basis for energy planning based on existing validated energy tools.

The first difficulty encountered was related to **the lack of relevant geometric that has an impact on the energy results** in public databases. Estimations on the dimensions of openings, lengths of thermal bridges, etc. had to be applied according to the types of buildings so as to consider these aspects in the energy simulations.

Also the work with **building typologies was necessary in order to apply thermal characteristics** to the building elements considered. This assignation was based on the climatic zone where the building is located as well as its year of construction.

Moreover, even when these assumptions were implemented in the platform, issues arose with respect to the **geometric definition of the buildings encountered in the Spanish Cadastre**, where not all the information was homogeneously defined.

With respect to the **calculation methodologies** implemented, several stumbling blocks were encountered. Firstly, the analysis of the Energy Performance Certificate tools showed great differences among each other, mainly in the input data process. Once CE3X was chosen, the file formats used in this tool presented a complex structure, which was difficult to work with since the elements (walls, openings, or other type of data) reflected in the files were not easily identifiable.

Furthermore, working with the tool when considering a restricted area (limited number of buildings) was manageable, but **when applying the approach at a bigger scale (for instance, citywide), problems emerged at all steps of the process**: public sources did not allow querying such amount of data at the same time and the simulation time increased drastically, sometimes leading to the interruption of the process.

Thus, in addition of the validation of the results mentioned before, a closer look to the improvement of the working process was necessary.

However, regardless of the difficulties described above, having analysed citywide data, it was shown that this platform can have a high potential in the support of energy planning activities within local authorities.

VII. CONCLUSIONS AND FUTURE WORK

The ENERGIS platform combines existing public data, validated calculated methodologies and GIS capabilities to offer a complete and powerful product to support energy planning.

The process to define, design and develop the platform has followed several stages, which involved the analysis of data coming from public sources, working with EPC tools, the validation of results at all stages and working with GIS in order to explore its full potential. Difficulties in this process involved mainly the work with public sources of data, in particular the Spanish Cadastre. The main problem was the lack of a standardised approach in the definition of geometric data. This fact would be solved with an adequate implementation of the INSPIRE Directive, which would guarantee homogeneous information of the buildings.

The platform provides users energy planning capabilities which will aid in the decision-making process when generating energy plans, by providing with maps showing the energy demand of cities and urban settings. It will reduce uncertainties and provide a knowledge base upon which to ground decisions.

Future work related with this platform includes: (1) enhancing the displayed data (providing consumption estimation), (2) performing calculations at dwelling level, or (3) offering new functionalities, such as capacity to introduce improvement measures.

All in all, by supporting energy planning, the adequate allocation of resources will be fostered to comply with energy directives' objectives and thus, energy retrofitting interventions will be boosted. As a consequence, the energy consumption of buildings will be reduced, resulting in lower emission rates, which will finally contribute in the fight against climate change, advocated by the United Nations with the signature of the Paris agreement.

ACKNOWLEDGMENT

This platform was developed in the framework of the collaborative project 'ENERGIS' in a consortium formed by Fundación CARTIF, VOXEL 3D and SYLTEC and cofounded by the *Agencia de Innovación, Financiación e Internacionalización Empresarial* and FEDER funds.

REFERENCES

[1] Paris agreement (2015) (COP 21): http://unfccc.int/paris_agreement/items/9485.php [retrieved: April, 2018]

[2] Communication COM/2016/0860 final. "Communication from the Commission to the European Parliament, the Council, the European Economic and Social Committee, the Committee of the Regions and the European Investment – Bank. Clean Energy for all Europeans". Brussels, 30.11.2016: http://eur-lex.europa.eu/legal-content/EN/TXT/?uri=CELEX:52016DC0860 [retrieved: April, 2018]

[3] Directive 2010/31/EU (2010). European Parliament and the Council of 19 May 2010 "On the Energy Performance of Buildings": http://eur-lex.europa.eu/legal-content/EN/ALL/?uri=celex:32010L0031 [retrieved: April, 2018]

[4] INSPIRE Knowledge Base: Infrastructure for spatial information in Europe: https://inspire.ec.europa.eu/ [retrieved: April, 2018]

[5] INSPIRE Roadmap: https://inspire.ec.europa.eu/road-map-graphic/32443 [retrieved: April 2018]

[6] New York City Map – City-Wide GIS.: http://maps.nyc.gov/doitt/nycitymap/ [retrieved: April 2018]

[7] Estimated Total Annual Building Energy Consumption at the Block and Lot Level for NYC: http://qsel.columbia.edu/nycenergy/ [retrieved: April, 2018]

[8] All buildings in the Netherlands: http://code.waag.org/buildings/#52.3884,4.9438,11 [retrieved: April, 2018]

[9] Energie Label Atlas (Netherlands): http://energielabelatlas.nl/# [retrieved: April, 2018]

[10] Spanish Cadastre Official Website : https://www.sedecatastro.gob.es/ [retrieved: April, 2018]

[11] B. Howard, L. Parshall, J. Thompson, S. Hammer, J. Dickinson, V. Modi. "Spatial distribution of urban building energy consumption by end use". ELSEVIER Energy and Buildings 45. (2012) pp. 141- 151.

[12] M. Kavgic; A. Mavrogianni, D. Mumovic, A. Summerfield, Z. Stevanovic, M. Djurovic-Petrovic. "A review of bottom-up building stock models for energy consumption in the residential sector". ELSEVIER. Building and Environment 45 (2010) pp 1683 – 1697.

[13] I. Miranda Pereira, E. Sad de Assis. "Urban energy consumption mapping for energy management". ELSEVIER Energy Policy 59. (2013) pp. 257- 269.

[14] R. Nouvel, M. Zirak, H. Dastageeri, V. Coors, U. Eicker. "Urban energy analysis based on 3d city model for national scale applications". IBPSA BAUSIM 2014 - Fifth German-Austrian IBPSA Conference RWTH Aachen University. 2014

[15] Spanish Cadastre Official Website. https://www.sedecatastro.gob.es/. [retrieved: April 2018]

[16] Spanish National Building Code (Código Tenico de la Edificación). Documento Básico HE: Ahorro de la Energía (Document on Energy Savings) : https://www.codigotecnico.org/images/stories/pdf/ahorroEnergia/DBHE.pdf [retrieved: April 2018]

[17] Energy Perfomance Labelling Technical Basis Manual of existing buildings CE3X. (Manual de Fundamentos Técnicos de calificación energética de edificios existentes CE3X), IDAE, 2012. http://www.minetad.gob.es/energia/desarrollo/EficienciaEnergetica/CertificacionEnergetica/DocumentosReconocidos/Documents/Manual_fundamentos_tecnicos_CE3X_05.pdf [retrieved: April 2018]

[18] E. Olivero, E. Onillon, P. Beguery, R. Brunet, S. Marat, M. Azar. "On key parameters influencing building energy performance". Proceedings of BS2015: 14th Conference of International Building Performance Simulation Association, Hyderabad, India, Dec 7-9., 2015

Situation Analysis of Electric Vehicles, Renewable Energy and Smart Grids for Sustainable Urban Mobility in Five European Regions

Kamilė Petrauskienė, Jolanta Dvarionienė,
Daina Kliaugaitė, Giedrius Kaveckis
Institute of Environmental Engineering
Kaunas University of Technology
Kaunas, Lithuania
e-mail: kamile.petrauskiene@ktu.edu
jolanta.dvarioniene@ktu.lt

Julie Chenadec
Green IT Amsterdam Region
Amsterdam, the Netherlands

Leonie Hehn
Barcelona Chamber of Commerce
Barcelona, Spain

***Abstract*—Urban electric mobility and renewable energies are innovative and smart technologies that can significantly contribute to reducing greenhouse gas emissions and increase energy efficiency. In order to increase their integration, the INTERREG EUROPE "EV ENERGY" project was initiated by five European countries (Italy, Lithuania, Spain, Sweden and the Netherlands). The project partners shared their mobility policies in order to learn how to overcome obstacles for the implementation of sustainable urban mobility. This research project comprises a situational SWOT analysis of electric vehicles, renewable energy and information and communication technologies. The greatest weakness that most countries encounter is the high price of electric vehicles and the lack of a developed charging infrastructure, however, a high number of private initiatives and an increasing people's awareness are promising.**

Keywords-electric vehicles; renewable energy; smart grids; mobility; situation analysis.

I. INTRODUCTION

Transport systems have a significant impact on the environment, accounting for 33.1 percent of the total energy consumption in 2015 [1]. Furthermore, the transport sector has the biggest oil demand at the European level, as two-thirds of the final demand for oil comes from transport. Road transport accounted for 54 percent of the final demand for petroleum products in 2014 [2].

According to Eurostat, all the sectors, except fuel combustion in transport and international aviation, contributed to the overall greenhouse gas emission reductions from 1990 to 2015. In 2015, transport emission rose for the second consecutive year. Besides, the average share of energy from renewable sources in transport increased from 1.4 percent in 2004 to 6.7 percent in 2015 [3].

Electrification of road transport has become a major trend and two of the most important technologies in European countries are Electric Vehicles (EVs) and Renewable Energies (REs). There is a spectrum of EVs technology, such as hybrid electric vehicles, battery electric vehicles and plug-in hybrid electric vehicles [4]. EVs can significantly reduce global and local emissions and are part of the future vision for global mobility.

Additionally, EVs are not only used for mobility, but also for energy services, including: energy back-up services, EVs participating in energy market, vehicle delivering energy to home/street/grid/city applications, variation in EV charging connected to local energy sources, accelerated storage of electricity in EV and/or electric public transport, implementing renewable energy sources for EV and/or electric public transport [5].

In the framework of the INTERREG EUROPE "EV ENERGY" project, an innovative approach was used to develop a framework of policies and experiences where energy, mobility and smart grids will work together. The project consists of partners from five European countries: Italy, Lithuania, Spain, Sweden and the Netherlands. The partners recognized a need to share technical and practical experience to learn how other institutions overcome the difficulties for the implementation of sustainable mobility. Interregional exchange and cooperation is necessary in order to achieve a better framework of policies, making it possible to develop more sustainable cities [5].

The main objective of this paper is to conduct the situation analysis of EVs, RE and Information and Communication Technologies (ICT) for sustainable urban mobility in five European regions.

II. METHODOLOGY

The study area of the current research comprises five European regions: Italy (Lazio region), Lithuania, Spain (Barcelona), Sweden (Stockholm) and the Netherlands (Amsterdam and Flevoland). The situation analysis was performed in each project partners' region by using Strengths, Weaknesses, Opportunities and Threats (SWOT) matrix, based on the inventory of good practices and mobility policies in each region. All the partners were actively involved in the data collection process. To ensure harmonisation of information and data collection, templates for the inventory of good practices and mobility policies were prepared and spread among the partners. The good practices and mobility policies were analysed on the city/regional, national and the EU scale. Additionally, the inventory was based on the three pillars approach 1) EVs; 2) REs; 3) ICT. Information and data were obtained by analysing secondary resources: bibliography, laws/regulations, relevant sites, public reports, national and international standards. In addition, focus groups meetings gathering policy makers, researchers and business representatives were held in each project partner's countries. What was collected within the inventory of good practices and policies contributed to the SWOT analysis.

III. Results and discussion

This part of the paper presents features of EVs, renewable energy and smart grid-ICT sectors integration as SWOT analysis in 5 European regions. The most important features are presented in Table I.

TABLE I. MAIN FEATURES OF SWOT ANALYSIS IN FIVE REGIONS [5]

	STRENGTHS	WEAKNESSES	OPPORTUNITIES	THREATS
Italy	- Kyoto rotation fund grants 50 million euro to support sustainable mobility. - Exclusion of EVs users from motor vehicle tax for 5 years. - Increasing production of renewable energy.	- High purchase price of EVs and their batteries. - Government lacks a clear legislative basis and national/ regional financial instruments. - High motorization rate in Rome (9 vehicles per 10 inhabitants) causes high pollution, noise and congestion.	- According to new construction regulations, new housing must have e-charging infrastructure. - Government initiated the National Infrastructure Plan to develop a recharging system in Italy.	- Most of e-charging network is not developed. - Strong dependence on traditional energy sources (oil products)
Lithuania	- Vilnius city allowed EVs to use bus and taxi lanes. - Parking EVs is free of charge.	- High purchase price of EVs and their batteries. - Lack of vehicle taxation mechanisms encourages people to use second-hand vehicles.	- Rejuvenation of a national car pool and reduction of dependence on oil products. - Opportunities for business and science in developing a new market and increasing car sharing competitiveness, additional subsidies and support mechanisms.	- Political decisions are lagging and are a way behind other EU states. - No single body responsible for e-charging infrastructure development. - High risk that second-hand EVs will be more expensive than fuel powered cars.
Spain	- Financial support for the development of e-charging infrastructure. - Lots of sunny days in Spain - a great potential of photovoltaic solar electricity.	- E-charging infrastructure lacks developed electrical networks and corridors, adequate power supply and maintenance that is more consistent. - EVs and hybrids take only 0.4 percent of the market.	- Growing investments and decrease of the photovoltaic energy price increase the number of EVs owners. - Promising initiatives of e-charging infrastructure, lowering Barcelona's energy dependence on imported energy.	- Lack of facilitating regulations, standards, training and framework for implementation, various barriers for public access to e-charging stations. - Limited control mechanisms to store surplus energy and generate additional during high demand.
Sweden	- Subsidies and tax reduction for EVs. - The power grid used for e-charging network is more or less already developed and is powered by mostly renewable energy.	- High purchase price of EVs and their batteries. - Companies expect high return, and the value of used EVs is low.	- Good potential of light EV. - The costs of EVs can be shared between partners. - People are eager to produce their own energy and use it on their own mobility.	- Additional tax if charging at the workplace and no parking benefits. - Budgets of municipalities too small to finance high costs of EVs and their infrastructure.
The Netherlands	- Part of MRA Elektrisch programme (20 000 EVs by 2020). - Culture of close public-private cooperation.	- Complexity of integrated energy projects. - Need of multi-stakeholder cooperation.	- Modern energy network. Spacious cities. - Awareness of need for integrated energy systems.	-Unstable national EV policies. - Large part of PV in rural areas. - Lack of awareness toward EVs. - Limited infrastructure.

IV. Conclusion

One of the main weaknesses is the high purchase price of EVs and their batteries. Another weakness is the lack of cooperation between stakeholders: municipalities, business enterprises, research and public bodies. Although some of the regions have already developed an e-charging network, further development is required. However, a developed e-charging network does not encourage people to change their ordinary fuel-powered vehicles, mainly because of the high EVs purchase price. On the other hand, countries are adjusting their policies and starting a number of initiatives, which might increase the use of EVs and integration of renewable energy, e-charging infrastructure and ICT.

Additionally, the use of EVs has another vision as energy storage. As REs have been increasing their weight in the energy mix year after year, EVs' batteries can be used to store locally produced RE, and give it back to the city during the hours of highest demand. This shows how the addition of EV to a household equipped with solar panels can greatly reduce the reliance on the grid. Such and similar cases present good practices and innovative policy making through regions, help other cities to reduce greenhouse gas emissions and save money in mobility and energy systems.

Acknowledgment

This work is a by-product of research during our work in the project "EV ENERGY" – "Electric Vehicles for City Renewable Energy Supply" (Project no: PGI02049). The project is financed by the INTERREG EUROPE programme (European Regional Development Fund).

References

[1] EUROSTAT. Statistics explained, "Consumption of energy," 2017. [Online]. Available:http://ec.europa.eu/eurostat/statistics-explained/index.php/consumption_of_energy. [Accessed: 10-Jan-2018].

[2] L. Buffet, "Europe increasingly dependent on risky oil imports," *Transport & Environment*, 2016. [Online]. Available:https://www.transportenvironment.org/sites/te/files/publications/2016_07_Briefing_Europe_increasingly_dependent_risky_oil_FINAL_0.pdf. [Accessed: 14-Jan-2018].

[3] EUROSTAT. Statistics explained, "Europe 2020 indicators – climate change and energy," 2017. [Online]. Available: http://ec.europa.eu/eurostat/statistics-explained/index.php/Europe_2020_indicators_-_climate_change_and_energy#cite_note-1. [Accessed: 14-Jan-2018].

[4] L. Raslavičius et al., "Electric vehicles challenges and opportunities: Lithuanian review," *Renewable and Sustainable Energy Reviews*, vol. 42. pp. 786–800, 2015.

[5] INTERREG EUROPE Programme Regional Initiative Project "Electric Vehicles for City Renewable Energy Supply". Project No: PGI02049 EV Energy (Duration January 2017 – June 2021) <https://www.interregeurope.eu/evenergy/>; 2017

PLECO: New Energy-Aware Programming Languages and Eco-Systems for the Internet of Things

Jon Robinson Kevin Lee
Department of Computing & Technology
Nottingham Trent University
Nottingham, UK
e-mail: jon.robinson@ntu.ac.uk
e-mail: kevin.lee@ntu.ac.uk

Kofi Appiah
Department of Computing
Sheffield Hallam University
Sheffield, UK
e-mail: k.e.appiah@shu.ac.uk

Abstract— **This paper outlines the aims of the Programming Language ECO-system (PLECO) to create new energy-aware programming languages and eco-systems for the Internet of Things (IoT). It builds upon the Lantern language and focuses on energy-awareness, security, resilience and communications for the large infrastructure underpinning the next generation of IoT. The paper outlines how IoT applications and deployments need to be developed in an energy-aware, secure and cost-effective manner using new secure, robust and energy-focused programming languages and the importance of taking such an approach.**

Keywords-energy-aware; Internet of Things; programming; distributed computing; security; Cyber-physical systems.

I. INTRODUCTION

It is projected that more than 50 billion Internet enabled devices will be online within the next 10 years [15]. This poses a problem for current ways of developing Internet of Things (IoT) and Cyber-Physical Systems (CPS) software as current practices do not consider the energy expenditure that these devices will introduce on existing power distribution networks. At present, developing applications for the IoT/CPS exposes devices to a number issues relating to energy use, security and reliability. IoT applications are currently developed using existing languages, frameworks and toolkits [16], which, in the case of programming languages, have not altered since their initial creation. Dynamic scripting languages like Python and JavaScript are being embraced by most IoT/CPS designers at the expense of high runtime cost due to the dynamic types and code optimisation techniques (e.g., Just-In-Time compilation) [19]. The applications they produce will tend to be less efficient and insecure [17] as the underlying development approach and programming language was initially designed without considering the core concepts of resiliency, energy-awareness and security. This leads to these concepts being added as an afterthought rather than as the primary focus of well-engineered software systems.

To engineer these applications appropriately requires the concepts of resiliency, energy-awareness and security to be central in the design and implementation of a system. To enable this, it is proposed that a new development approach, built on a language focused around these concepts be the way in which systems are written. This would ensure that software is secure, reliant (i.e., dealing with communication, distributed complexity and failover) and importantly, energy-aware from their inception by allowing developers to implement them using algorithms which promote these areas.

In Section II, the need for energy-awareness is discussed and current approaches to Internet of Things software development is introduced. In Section III, the proposed PLECO architecture is introduced and discussed. In Section IV, the experiences learnt from the initial Lantern energy-aware domain specific language is discussed. Finally, in Section V, a summary of the work is provided.

II. CURRENT PRACTICES

Energy-efficiency is a growing research focus in all areas of technology, including IoT. Energy-utilisation of hardware had been addressed widely in the embedded systems area where software can reduce the power usage of components of the underlying hardware. However, energy-awareness is still poorly represented when it comes to building large scale distributed systems and the algorithms used to implement them [18]. Additionally, languages suffer from providing developers with practices and language constructs which have been available in general purpose languages for many years. However, when these languages were initially designed, the computing and distributed landscape was significantly different than from what it is today. Concepts representing how to use energy-aware algorithms for efficient interacting distributed systems, coordinate, adapt, self-heal, secure and be resilient have not been fully considered in their design phases.

Energy-awareness has been a major focus within the embedded systems world where the conservation of energy is instrumental in the operation of a device. Incorporating energy efficiency within the design of the circuitry underlying the device has proven to be effective [1] and acts as another justification on the PLECO approach. Other static approaches to embedded design have been proposed [2][3].

Energy-awareness in software development has highlighted many challenges in the production of energy-aware software systems. For example, application-level approaches advocate applications being energy-aware and controlling their own energy use [4]. To support application development, tools that monitor applications to provide energy use information [5][6] exist to aid in the process. In addition, by incorporating middleware to help with efficient energy usage in applications without them explicitly being aware of this focus [7][8] could be utilised. The need for a dedicated programming language with resource constraints to streamline usage was identified in [9] and has been a research trend mainly for security [10]. There are other approaches to IoT development which fall into cloud, Operating Systems, middleware or protocols (e.g., IF This Then That (IFTTT), Azure IoT or AWS IoT, Kontiki, Brillo), MQTT, Gaia, etc.). In [11], updates to the International Technology Roadmap for Semiconductors (ITRS) [12] are discussed. The ITRS provides a roadmap of hardware and software technologies in the design and development of silicon systems. The road map outlines the trends of future technologies to address challenges regarding the cost of design and power / energy use. Future trends within the ITRS show that power-aware systems are currently a challenge in the control of electronic devices. Thus, with the popularity of the uptake of IoT devices, programming them in an energy-aware manner is an important problem to address.

III. THE PLECO ARCHITECTURE

Current software development paradigms are not ideally suited to tackle the energy efficient and distributed nature of IoT and similar technologies. This is exacerbated by the lack of energy, security and reliability standards and frameworks for this domain. This opens up the need for alternative methods for developing, deploying and supporting software for IoT deployments and other related applications.

To solve this problem, a shift in software development which enables the efficient design, development, support systems and eco-system for these emerging distributed systems technologies. By focusing on the principles of low energy, security and reliability, this will best serve the needs of large-scale heterogeneous systems. We propose the development of a complete eco-system for modern software development, including development languages and support systems which enable the efficient design and use of system wide energy, and production of software systems which address the key challenges of modern Internet based systems.

The PLECO architecture aims to investigate new paradigms and languages for software development in large scale distributed systems by introducing a new energy-aware, reliable and security focused eco-system. This aligns with the ITRS future goal of energy-aware programming for IoT.

We advocate the there are three fundamental pillars underlying the design and development of a new eco-system for developing energy-aware systems. For each of these pillars, by directly integrating them into the development process, developers would plan, design, implement, test and deploy applications which satisfy these programming styles from the onset: Energy-Awareness, Security and Reliability.

The PLECO system builds on preliminary work into energy-aware Domain Specific programming languages and middlewares [13][14]. With the growing acceptance and use of IoT enabled devices and the lack of security with these devices, a new way in which to design, construct and implement solutions needs to be considered. At present, there are very few frameworks which directly address the robust production of IoT systems. However, security is only a secondary concern which leaves devices open to exploitation. Thusly, the adoption of a new way in which to design and build large scalable, secure, and robust distributed systems requires a new platform and language is required.

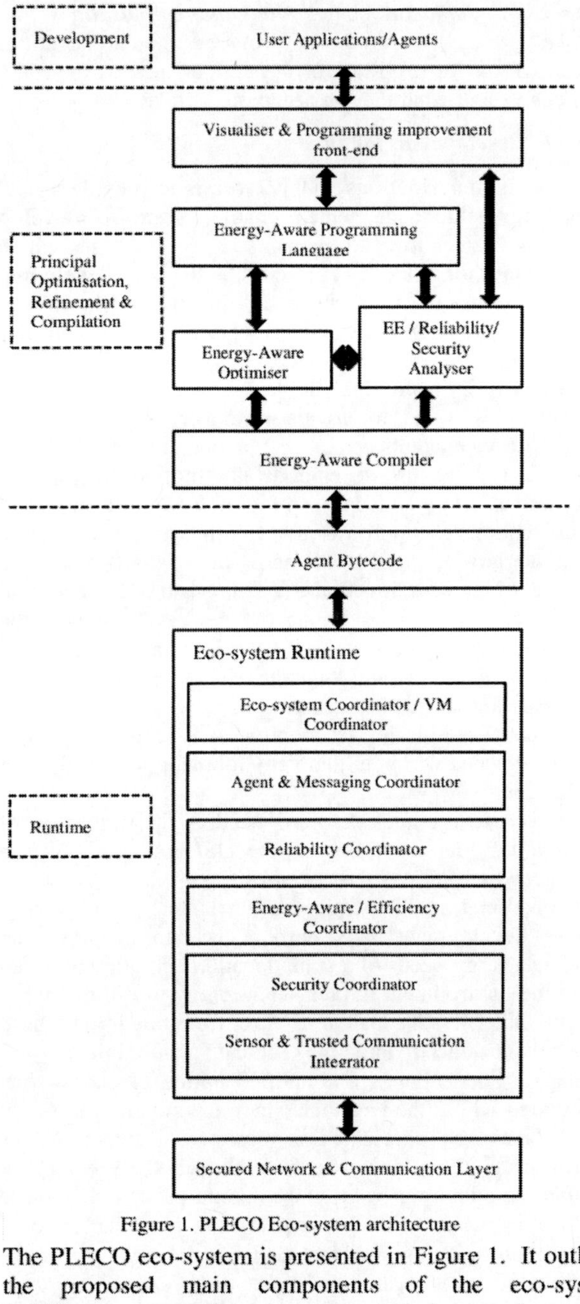

Figure 1. PLECO Eco-system architecture

The PLECO eco-system is presented in Figure 1. It outlines the proposed main components of the eco-system (middleware, compilers, language, optimisers and potential standards). This builds on previous work into service composition middlewares [13] and energy-aware domain specific languages (Lantern) [14].

The main contribution of this work can be summarised as: 1) fully complete energy-aware programming language for controlling and managing IoT devices; 2) inclusion of security into language design for secure software development as well as failover support; 3) distributed concepts for the management and communications in highly scalable IoT architectures; and, 4) infrastructures for supporting new methods. The approach taken is to consider the development and runtime in three distinct phases. Namely, *development, compilation and optimisations* and *runtime*. What follows provides the architectural breakdown of our proposed eco-system as well as the main components and what their expectations will be.

Layer 1: Development language

One of the contributions of this work is to provide a new programming language which enables users to develop energy-aware, secure and robust software systems which builds upon prior work. At present, the focus on developing systems is to use existing languages which forces developers down specific design routes which requires them to consider energy usage and security as secondary considerations. The Lantern [14] Domain Specific Language (DSL) has been previously developed, to provide developers with a language in which service agents can be constructed and was designed from the ground up to embody the notion of energy-awareness as a key concept (see Figure 2 for an example of Lantern code). The purpose of the Lantern system was to investigate how to provide an energy-aware domain specific language which was aimed at managing and controlling the energy consumption of disparate IoT devices. This was the first test iteration of a language to test ideas and confirm the viability of energy-aware languages. These agents interface with hardware-based devices as well as providing energy-aware adaptive abilities to monitor and adapt the power usage of devices within a home environment populated with IoT devices.

However, as it stands, Lantern acted as the first stage of investigating adaptive energy-aware DSL's and provides a language which allows the control of devices rather than a fully functional and semantically rich language for general purpose development. Its purpose was to adapt to the changing energy needs of a static location and allow devices within the environment to alter their energy use, thus saving energy. Other lessons that were learnt from this initial phase will be introduced into the second generation of the language. For example, a simplified notion of energy was represented within the language where the amount of energy used (Watts) was represented as values associated to power control structures. However, even though these were not strongly typed power values, the intention was that these would infer the amount of energy used. No other form of energy representation was included but the next step is to represent energy in a variety of more strongly typed language constructs which could represent Joules, or other energy-based representations (e.g., temperature).

Hence, the purpose of the next generation of the language is to provide developers with a new way of developing software programs for Internet of Things devices. It will be a semantically rich language, rather than a DSL which will provide them with the ability to provide more sophisticated software. By providing a new language, rather than using an API, requires them to consider the energy-awareness of the design and operation of the software, security and robustness which can have a direct impact of the energy use of the smart environment they are located within. This ensures that the considerations and requirements of writing software for today's highly distributed systems are considered from the outset of development rather than as a secondary consideration. For example, C and C++ languages have been in use for decades before large scale distributed networks of cooperating IoT devices where considered. Because of this, the underlying languages do not provide the concepts of security, energy-awareness and robust as central tenets of the language and, hence, at best are considered after the design and during the implementation of systems, and often not at all.

```
aliases {
  alias( heating_control ) -> device( heating, ERD204)
  alias( heating_temp ) <- device( heating, ERD204)
  alias( motion_control ) -> device( motion, ERD204)
  alias( lights_control ) -> device( lights_1, ERD204)
  alias( lights_power ) <- device( lights_1, ERD204)
  alias( PC_control ) -> device( pc1, ERD204)
  alias( PC_power ) <- device( pc1, ERD204)
}
environment(ERD):{
  location(ERD204) : {
    uses device( heating ) <- input(heating_temp)
    uses device( heating ) -> output(heating_control)
    uses device( motion ) <- input(motion_control)
    uses device( lights ) <- input(lights_power)
    uses device( lights ) -> output(lights_control)
    uses device( PC ) <- input(PC_power)
    uses device( PC ) -> output(PC_control)
  }
}
consumption(ERD):{
  override( ERD204 > 800 ) -> {
    condition( heating > high ) -> action( heating = off )
    condition( lights == on ) and condition( !movement ) -> action( lights =
off )
    condition( PC == on ) and condition( !movement ) -> action( PC = off )
  }
}
(identity:cmp3robinj):(location:ERD204) {
  condition( at(7:30) ) -> action( heating = on )
  condition( temperature < low ) -> action( heating = on )
  condition( temperature > high ) -> action( heating = off )
  condition( lights == off ) and condition ( !movement ) -> action( lights =
on)
  condition( at(20:00) ) and condition( !movement ) -> action( lights = off )
  condition( at(20:05) ) and condition( PC == on ) and condition(
!movement ) -> action( PC = off )
}
```

Figure 2. Example Lantern code

This will introduce new ways to represent the energy-awareness of systems through language constructs which enable devices and software systems to be actively aware and adapt their power utilisation. For instance, rather than focus on energy consumption within the hardware level, energy-aware constructs will allow software to be built which is both efficient and energy-aware by using algorithms and program design which facilitates in reducing the overall energy expenditure of the interacting system. The notion of security is currently poorly represented in software design, so another key area of the language is to incorporate secure development and language constructs from the outset. This will enable developers to consider security related considerations in the design and implementation of systems by using algorithms and language constructs, which promote secure systems. The final component of the language is to incorporate constructs which allow resiliency (and robustness) within interactive systems. The complexities of distributed systems also will be addressed by providing distributed management constructs. This is to ensure that systems can adapt and reconfigure themselves if components of the larger system fail or are unavailable.

Layer 2: Optimising technologies

The purpose of this layer of the eco-system is to provide programming support to developers and users. A variety of sub-systems will be provided, which allow for the analysis and improvement of software by enabling support for code optimisation. A number of key components are required:

Energy-aware programming language: As has been previously discussed, the language will offer the concepts of energy-awareness (by allowing the monitoring and adaptation of energy use within an environment), resiliency/robustness through failover, distributed complexity, communications and management/control; and security through secure communication and language constructs. It will provide support on how to write adaptive systems which can react according to environmental stimuli to make best use of the resources on offer depending on the energy requirements of devices. A focus on the representing the interactions between devices and associating an energy cost to these interactions will enable algorithms to start to consider the economics between device interactions. The language will be extensible and introduce notions of *low carbon foot-printing, identity, distributed systems, agents, data generation, composition, mobility, reconfiguration, security, privacy, trust* built in which provides users with the means for writing effective systems.

Energy-aware optimiser: Optimisation of energy-aware systems requires the analysis of both the programming style and algorithms used within the software construction stage and the way in which agents will interact with each other to make best use of the resources that are on offer. This will primarily focus on inspecting the code written by the end-user to analyse whether there are more efficient ways of representing the code which can be made. It will not analyse how to make efficient use of the underlying hardware (e.g., turning off Wi-Fi, controlling processor state, etc) but instead will examine the algorithms that have been used to see if it is possible to increase their energy efficiency by modifying how they work and how they interact within a larger system.

Energy-Aware/Reliability/Security Analyser: This component provides analysis of the agent based on the notions of reliability and security. It will determine whether the best practices have been followed to ensure that the agent is secure. It will also analyse the agent to determine if it is reliable and robust (i.e. distributed complexity as well as securely constructed, including secure communication). Debugging information will be generated which allows higher-levels to visualise data based on how to improve the security and debugging of agents within the system.

Energy-Aware compiler: The energy aware compiler will produce byte-code which makes best use of the three main concepts behind the language to generate agents. Generated byte-code will be executed within a safe, secure and reliable environment provided by the eco-system runtime in line with the existing Lantern system. This is currently being developed and offers agents a distributed playpen in which to execute. The agents that are produced will automatically bind to the runtime which offers a controlled exposure to the underlying runtime properties. Performance and profiling information which includes programming specifications outlining the type of data produced, consumed and linked to, to help with the generation of mobile and location aware agents will be considered. The compiler would optimise applications based on the corpus of data generated by the user and how the user intends the agent to interact within the eco-system and how it consumes data and its reliance on other agents. The compiler will determine the appropriate hardware requirements and locality (i.e., closeness to work) where agents running within the runtime close to where tasks needs to be completed.

Visualisation & Programming style improvement system: Another key way in which languages need to be supported is through verification, validation and visualisation of software systems. This will provide a graphical front-end for improving agent design and development. A hints/help system will provide the user with ways in which to increase system efficiency by suggesting improvements to security and reliability considerations.

Layer 3: Eco-system runtime

The purpose of the runtime system is to provide a consistent environment for executing agents. It is comprised of many sub-components that provide control for: coordination; discovery; invocation; virtualisation; agent mobility; reliability (i.e., failover, tolerance); energy-awareness and adaptation; distributed complexity; and, trusted communication & security. This builds on the existing Lantern middleware as well as other areas that are currently being investigated. The ecosystem runtime layer will be formed out of the following coordinator sub-components:

Ecosystem coordinator / VM coordinator: This will provide executing agents a safe, protected, virtual environment to run within. Mobility of agents will be managed within this level so that they can make use of the resources within the environment. This was an issue with Lantern as the language and agent were based on statically located devices. For this iteration, agents will be mobile and be able to transport themselves around within the eco-system. This will mean that location dependency will ensure that agents are running and interfacing with the best set of devices depending on the whereabouts of users and deployed system. Exposure to adaptive and reconfigurable aspects of the runtime will be provided to agents so that they are able to locate and adapt themselves to their surroundings. Agents will be exposed to the discovery and linkage to other agents provided by the agent and messaging coordinator. Therefore, this will provide agents with a distributed, reconfigurable, compositional and collaborative runtime for the coordination and secure control of systems. The runtime will also provide identity management and identity conflict resolution. This builds on the initial Lantern representation of identity which was weakly defined. For this iteration, to help with managing a number of identities, a group based approach will be taken which allows identities to inherit permissions and access control for different environments.

Agent & Messaging coordinator: This will maintain agents by providing them with resources within the runtime and provide them with mobility facilities. Control of processing resources, complexity, memory, storage, and message handling will be offered. The adaption and reconfigurable nature of agents will be handled by this coordinator so that they can adapt to conditions over time.

Reliability coordinator: This will ensure that agents can deal with situations where something goes wrong. This will be through a combination of approaches ranging from agent reconfiguration; failover control; re-incarnation; complexity; agent adaptation (where agents can self-heal); and, debugging information and mechanisms for diagnosing interaction and programming issues.

Energy-aware/efficiency coordinator: This will coordinate the most efficient use of devices and interactions / collaborations with other agents. Its primary purpose is to provide exposure to the energy consumption aspects of interfacing technology (i.e., actuators connected to devices). It will also control the re-configurability and binding of agents to devices based on power needs.

Security coordinator: This will provide the underlying security model for managing and coordinating agents and devices. Secure and trusted communication between devices will be provided for. It will offer protection from tampering of agents and devices from malicious entities (e.g., other users and systems) and provide cyber-security attack resistance.

Sensor & Trusted communication integrator: This will provide exposure to an "Internet of Trust" layer for which facilitates the trusted communication between agents to guarantee secure and private communication within the eco-system. Working in conjunction with the security coordinator, it will formulate trust-relationships between independent nodes within the eco-system. This is used in the building of an "Internet of Trust" between devices running the eco-system and sensors, and the brokering of collaborations and transmission of trusted information between trusted content producers and content consumers. It will also determine how trust can be integrated into communication protocols in a bid to aid in the routing of information between trusted parties. The sensor integration aspect to this layer will enable devices running the eco-system to securely interface with agents.

IV. PHASE 1: LANTERN LESSONS

The PLECO language will expand on the initial development phase of this work. The first phase, was the design and development of the Lantern Domain Specific Language (DSL) [14]. The key aim of this was to test ideas on how energy-awareness could be represented within programming language design for controlling Internet of Things based devices within a home environment. Based on these findings, several lessons were learnt which will be built upon for PLECO.

Figure 3a provides a hierarchical overview of the language constructs available within Lantern, while in Figure 3b shows the key components of the Lantern middleware.

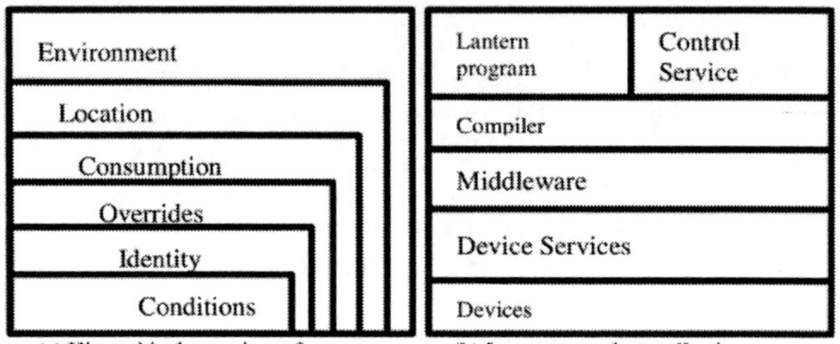

(a) Hierarchical overview of language constructs

(b) Language and compiler / middleware stack

Figure 3. Lantern overview

An example of the Lantern language is given in Figure 2. The language took the approach of providing developers with a DSL which allowed environments to be represented within the language. These environments would in the real-world, translate into building representations. Any number of environments could be provided and allowed the definition of statically defined locations to be provided. These locations allowed *environments* to be divided up into smaller administrative boundaries to represent rooms, hallways and other types of locations and provided a means in which the grouping and control of devices located within these statically defined areas could be provided. Consumption rules provided the user with a way of specifying a number of rules in which to monitor the energy usage within these locations and provided a means in which the environment would adapt and use devices located within the location. The *consumption* construct acted as a container object where all rules which needed to override user specific energy utilisation rules would occur. The *override* construct would be used in tandem with the consumption definition to provide the fine grained adaptive nature of the environment. *Identities* were weakly defined within Lantern and represented the user based on their name or identifier. This meant that to allow the environment to be truly adaptive to various users, many identities would have to be defined. Because of this, a group-based membership approach will be used to coordinate this process. *Conditions* were used to represent the user defined rules governing the energy usage and adaptive nature of the system. An outline of the principal parts of a Lantern agent is shown in Figure 2.

As Lantern was to explore and experiment with languages to represent energy-awareness, it did not provide a platform for testing secure programming and communication styles or reliable software. However, several things were learnt from this initial phase which will inform PLECO. They are:

- using environments and sub locations proved to be quite effective in representing buildings and static locations. However, the mobility of devices and sensors was deemed to be insufficient to cope with environments which are dynamic and change over time. This will be investigated further so that the necessary coordination and control constructs are considered to allow situations where mobility is needed.
- telemetry constructs for allowing the flow of information from devices, as well as coordinating the control of such devices proved to be quite effective.
- an expanded set of strongly typed constructs for the representation of energy.
- a weak notion of identity was provided within Lantern and was not powerful enough to represent the number of users within an environment which resulted in scalability issues. To address this, a group-based membership/user approach to allow the inheritance of security principals as well as the management of group-based identities will be adopted.
- a security concern in the Lantern language showed that identities could be mapped to individuals due to the simplistic way in which identities were programmed (string based). However, this will be expanded upon to provide anonymous identities as well as obfuscation of user identities.
- condition rules allowed the introduction of the notion of time (i.e., at a point in time, do something). This will be expanded upon to provide users with more time-based constructs to deal with different length durations (e.g., to reduce the energy use for a specific amount of time in a day).
- conditions provided a clause in which an action would be performed once something had happened. This was found to be adequate but improved structures and language constructs for handling more complex interactions and reacting to non-time bound interactions will be.
- to support verification, other approaches are being looked at (for example contract based and assumed guarantee reasoning).

V. CONCLUSION AND NEXT STEPS

This paper has introduced the PLECO architecture and its objectives. The eco-system to support the next generation of languages and middlewares which have been designed from the ground up to incorporate the notions of energy-awareness, security and reliability rather than added them as a secondary consideration. By incorporating these notions in the design and development of systems will provide more robust and secure systems in which to control large scale distributed IoT devices. The lessons learnt from the first iteration of energy-aware languages have provided a foundation on which to provide the language aspect of the eco-system.

REFERENCES

[1] S. Mittal. “A survey of techniques for improving energy efficiency in embedded computing systems”, International Journal of Computer Aided Engineering and Technology, 6(4), pp. 440–459, 2014.

[2] P. Yang, P. Marchal, et al. “Managing dynamic concurrent tasks in embedded real-time multimedia systems”, Proceedings of the 15th international symposium on System Synthesis, pp. 112–119, ACM, 2002.

[3] Y. Ma, N. Sang, W. Jiang, and L. Zhang. “Feedback-controlled security-aware and energy-efficient scheduling for real-time embedded systems”, Embedded and Multimedia Computing Technology and Service, pp. 255–268, Springer, 2012.

[4] F. Alessi, P. Thoman, G. Georgakoudis, T. Fahringer, and D. S. Nikolopoulos. “Application-level energy awareness for openmp”, International Workshop on OpenMP, pp. 219–232, Springer, 2015.

[5] N. Amsel and B. Tomlinson. “Green tracker: a tool for estimating the energy consumption of software”, CHI’10 Extended Abstracts on Human Factors in Computing Systems, pp. 3337–3342, ACM, 2010.

[6] M. Sabharwal, A. Agrawal, and G. Metri. “Enabling green it through energy-aware software”, IT Professional, 15(1), pp. 19–27, 2013.

[7] N. Nikzad, O. Chipara, and W. G. Griswold. “Ape: an annotation language and middleware for energy-efficient mobile application development”, Proceedings of the 36th International Conference on Software Engineering, pp. 515–526, ACM, 2014.

[8] Y. Xiao, R. S. Kalyanaraman, and A. Ylä-Jääski. “Middleware for energy-awareness in mobile devices”, Proceedings of the Fourth International ICST Conference on COMmunication System softWAre and middlewaRE, p. 13, ACM, 2009.

[9] D. Aspinall, S. Gilmore, M. Hofmann, D. Sannella, and I. Stark. Mobile Resource Guarantees for Smart Devices, pp. 1–26, Springer Berlin Heidelberg, Berlin, Heidelberg, 2005, URL https://doi.org/10.1007/978-3- 540-30569-9_1.

[10] D. Franzen. “Quantitative bounds on the security-critical resource consumption of javascript apps”, PhD Thesis, University of Edinburgh, 2016.

[11] G. Smith. “Updates of the itrs design cost and power models”, 2014 IEEE 32nd International Conference on Computer Design (ICCD), pp. 161–165, 2014.

[12] Semiconductor Industry Association. “The international technology roadmap for semiconductors”, URL http://www.itrs2.net, 2013. [retrieved: January, 2018]

[13] J. Robinson, I. Wakeman, and D. Chalmers. “Composing software services in the pervasive computing environment: Languages or apis?”, Pervasive and Mobile Computing, 4(4), pp. 481–505, 2008.

[14] J. Robinson, K. Lee, and K. Appiah. “Lantern – A Smart-Home Enabled Domain Specific Language for Energy Awareness in Cyber-Physical Systems”, Under review - unpublished, 2017.

[15] L. Ericsson. "More than 50 billion connected devices." *White Paper*, 2011.

[16] A. Al-Fuqaha, M. Guizani, M. Mohammadi, M. Aledhari, and M. Ayyash. “Internet of things: A survey on enabling technologies, protocols, and applications”, IEEE Communications Surveys & Tutorials, 17(4), pp. 2347-2376. 2015.

[17] Z. Zhang, et al. "IoT security: ongoing challenges and research opportunities", Service-Oriented Computing and Applications (SOCA), IEEE 7th International Conference on. IEEE, pp. 230—234. 2014.

[18] A. Orgerie, M. de Assuncao, and L. Lefevre. "A survey on techniques for improving the energy efficiency of large-scale distributed systems." ACM Computing Surveys (CSUR) 46.4 47. 2014.

[19] C. Kim, et al. "Typed Architectures: Architectural Support for Lightweight Scripting." Proc. of the 22nd International Conference on Architectural Support for Programming Languages and Operating Systems, pp. 77-90, ACM, 2017.

A Software-based Approach for Source-line Level Energy Estimates and Hardware Usage Accounting on Android

Alexandre Cornet and Anandha Gopalan
Department of Computing, Imperial College London, London, SW7 2AZ, UK
Email: alex.cornet.ac@gmail.com, a.gopalan@imperial.ac.uk

Abstract—**As users rely more on their mobile devices, energy inefficient software is a real threat to user experience. Early tools for developers focussed on expensive power measurement hardware and software-based approaches were introduced to relieve them of such requirements. These tools highlighted the most energy-inefficient parts of the code, but the developer still had to find and understand the exact causes of energy drain. Also, there was no mapping of hardware energy activity to code and no accounting for tail energy. To this end, this work focusses on providing source-line level energy estimates and maps the drain caused by Wi-Fi back to the code while accounting for tail-energy.**

Keywords–Green-computing, Tail-energy, Energy profiling

I. Motivation

Mobile technologies have become ever-present in our daily lives and many people's personal and professional interactions now depend on their smartphone or tablet. The power consumption of mobile devices obviously grew with the duration of their usage and the complexity of hardware and software they involve. By nature, these devices are used away from power sources and battery has thus become a critical component for the user experience. However, battery technology hasn't managed to keep up with these requirements and hence energy efficiency has become a major concern of users who are now looking for feedback to understand how applications drain the battery of their devices [1]. Research even shows that implementations perceived as energy greedy will receive lower ratings from users [2]. The ability to build energy efficient software consequently rewards developers with a competitive advantage on the market. However, energy optimisation is often counter intuitive to many developers and there is no real guidance. For example, there is no clear correlation between energy and time efficiency [3], and time optimisation is thus not always useful to reduce the energy footprint of a software. Also, some popular design principles, such as the decorator pattern have bad energy efficiency [4].

Researchers have consequently developed tools to provide guidance to developers by profiling the energy drain of their code. These tools were initially tied to specific and expensive power measurement platforms which inherently limited their use. Software-based approaches were introduced so that energy profiling techniques are accessible to the vast majority of developers by placing no hardware requirements on them. The first key contribution of this work is to provide source-line level energy estimates to the developers, which is achieved by extending Orka [5] [6].

To provide accurate and meaningful energy feedback, however, energy profilers should also take into account the drain caused by various hardware components. This means being able to map the hardware energy usage back to the code with the finest granularity possible. Moreover, mobile devices exhibit several asynchronous power behaviours, the most significant of which is *tail-energy*, which proves challenging for the development of energy profilers [7] [8]. A component exhibits tail-power behaviour if it stays in a high power for a constant time after processing a workflow and this phenomenon can cause significant energy drain. Thus far, only a few contributions, such as *eprof* [7] have focussed on tail-energy by relying on hardware- or model-based approaches. Therefore, no tool accounting for tail-energy was accessible to the majority of developers. To allow for this, the second major contribution of this work is for our solution to also provide tail-energy accounting. We have focussed exclusively on Wi-Fi for now and will in future include other hardware components.

The remainder of this paper is organised as follows: Section II outlines related work, Sections III and IV detail the major contributions. Section V provides an evaluation, and Section VI concludes the paper and provides ideas for future work.

II. Background Research

This section outlines other works related to this paper.

A. High level guidance

Software optimisation has tended to focus on time, but according to [3], choosing an energy-efficient implementation over a time-efficient implementation allowed more operations to be performed on a mobile device. [9] reviewed a set of coding practices and proved that reducing memory usage has a low impact on reducing the energy usage of code. Related conclusions were obtained in [10] as they proved code obfuscation negatively impacts the energy efficiency of the underlying software. In [7], they profiled and analysed the energy footprint of six of the top-ten applications on the *Google Play Store*. They found that most energy is spent in I/O and that most of this energy is due to tail-behaviour. The authors in [11] surveyed 55 applications in order to mine energy greedy calls to the Android API. They obtained the cost of 800 API calls and found that the Android API was the most energy inefficient part of applications. These findings are however, limited by the fact that part of the energy-greedy I/O activities are handled in the Java API and therefore don't appear in this study. Finally, tail-behaviour was not considered.

High level guidance provides a first mean to produce more energy-efficient software. However, these findings don't allow developers to know which specific parts of code consume the most energy. Researchers therefore focussed on developing energy profilers able to provide detailed feedback about the energy footprint of applications.

B. Hardware-based profilers

Hardware-based energy profilers such as *PowerScope* [12] and *GreenMiner* [13] involve the use of power measurement platforms, i.e., mobile devices with embedded power sensors monitoring the energy drain caused by running processes. *vLens*, an energy-profiler combining program analysis and statistical modelling to provide energy usage information at the source line level was introduced in [8]. These approaches however, are tied to expensive power measurement platforms, involving an overhead for their users.

C. Model-based profilers

Model-based solutions build power models of mobile devices, which are further reused away from measurement platforms to profile the energy footprint of applications. *eCalc* [14] assumes the availability of a CPU power profile containing the energy drain associated with each CPU instruction in order to provide drain caused by the CPU at the method granularity. *PowerBooter* uses battery voltage sensors and the knowledge of the battery behaviour to automate the power model generation and *PowerTutor* further uses this model to compute the energy costs [15]. *TailEnder* provides models focussing on the drain of I/O components, namely 3G and GSM. *eprof* [7] focusses on mapping the drain caused by hardware components while accounting for entities such as processes, threads and routines. For each entity, *eprof* produces an energy tuple `(utilisation_draw, tail_mode_draw)` and is therefore one of the few contributions accounting for tail energy and comparing it to utilisation drain. They also present methods to generate these power models by correlating certain behaviours with specific power states and identifying these behaviours as trigger conditions on the finite state machine [16]. Model-based techniques suffer one major drawback: the underlying power models are specific to the profiler and are therefore not widespread nor publicly available.

D. Software-based profilers

Software-based solutions were introduced to allow for portable and widely accessible energy profilers. They do not require any knowledge of the device behaviour or access to specific hardware. Users usually provide their application alongside a test scenario. Based on the findings in [11], [2] introduced *EcoDroid*, an energy profiler that focuses on ranking applications instead of providing energy estimates. The Power Estimation Tool for Android Application (*PETrA*) uses new tools of the Android Open Source project focussing specifically on energy profiling [17]. *PETrA* simulates a typical execution of the tested application on an actual device using a user-provided script. The execution trace is recorded using `dmtracedump` and the `batterystats` history is collected using `dumpsys`. *PETrA* finally replays these files to compute the battery drain during the execution of each method and provides the user with the energy estimates at the method level.

Orka is one of the first software-based energy profilers [5]. It takes an Android application and a dynamically created execution trace to provide energy usage estimations as feedback. The energy cost of a routine is calculated using the energy costs of the Android API calls (from [11]). We had access to the source code of *Orka* and its design and assumptions were compatible with our project goals. Hence, extending *Orka* proved to be the most viable solution. The basic workflow of *Orka*, shown in Figure 1, is divided into three parts: (i) application is instrumented to log the API calls, (ii) instrumented application is then run on the Android emulator using a user-supplied `monkeyrunner` script, and (iii) execution traces are analysed and results are presented to the user. The execution analysis is done by using the logs and `batterystats` data and is used to calculate the total cost of a routine which is obtaining using: $cost(routine) = \sum_{API \in routine} cost(API)$. The `batterystats` file also gives the breakdown by component of the hardware energy usage.

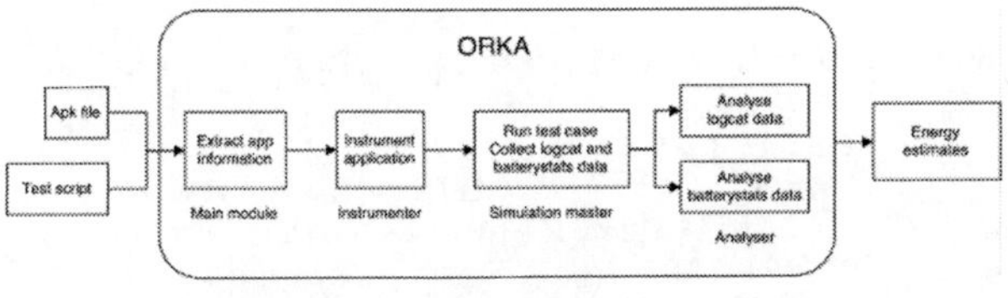

Figure 1. Basic workflow of *Orka*

III. PROVIDING SOURCE-LINE LEVEL ENERGY ESTIMATES

The key extension to Orka is to account for API calls occurring in subroutines and provide this at the source-line level rather than the method-level as was the case previously. To this end, *Orka* now maintains an in-memory version of the call stack while it replays the logs and attributes any API calls to all the routines in the call stack. Therefore, *Orka* tallies the calls occurring *during* the execution of this routine, possibly in a subroutine with the results displayed as a percentage of the total routine cost. Taking control-flow into account, the cost of a line of code is defined as the cost of a single execution of this line multiplied by the average number of times it was executed per call. *Orka* approximates the energy usage of a routine with the cost of its calls to the Android API and assumes that remaining parts of the code have a marginal energy footprint. On top of tallying the number of API calls during a routine execution, *Orka* uses the information provided by the `smali .line` instruction to know where in the source code each of these calls occurred. Since all the API calls happening during the execution of a subroutine are now attributed to the parent routine as well, for a given routine, all API calls happening during the execution of a subroutine will be attributed to the line from which the subroutine is invoked.

A. Implementation

1) Injector: The first step was to allow the instrumented application to log `(apiName, lineNumber)` pairs. On finding the `smali .line` statement, the injector now updates a variable containing the source line number corresponding to the current instruction. On finding an API invocation, this information is then appended to the API's name and passed to the `APILog` function. At this point, *Orka* is able to know where the API was called in a method's body and to detect when the API was called from a subroutine. In order to map this cost to a specific line in the parent routine, we have to allow the instrumented application to log messages indicating subroutine invocations alongside the line from which the subroutine was called. On finding an `invoke` statement, the injector now checks whether the called routine is user-defined and injects a call to a logging function if so. To enforce one main requirement of the injector, namely to ensure minimum code is injected, log messages shouldn't be inserted before the invocation non-injected routines, as this information won't be useful during the analysis. Therefore, the injector needs to know which routines are injected prior to the injection and add log messages only before calls to injected subroutines. To this end, a preparatory phase was added to the injector work-flow: injected files are now scanned prior to the injection in order to build the set of injected methods.

2) Analyser and `routine` class: The `routine` class calculates the cost of a routine as the total cost of all its API calls. To implement fine-grained energy feedback, we instead identify the cost of a routine as the sum cost of all its lines. The cost of a line is defined as the cost of the API calls occurring on

```
method RecList.onCreate, Average cost:
    0.0197754308, Calls: 1.0
      2.64% 1-1
      1.45% 1192      calling RecList.
        InitializeRepeatButton
      0.88% 1227      calling RecList.
        InitializeThemedButtonsBackgrounds
      0.57% 1229      calling Preferences.
        getCurrentlyPlayingFilePath
     29.04% 1247      calling RecList.
        updateSongList_extended
     21.80% 1257      calling android.os.Bundle.
        getString
     21.80% 1258      calling android.os.Bundle.
        getString
     21.80% 1260      calling android.os.Bundle.
        getString
```

Figure 2. Example of reconstructed source code with fine-grained guidance

that line. The `routine` class was thus extended to generate a reconstructed source code including energy estimates from the API data – typical output shown in Figure 2. To store the line from which a subroutine is called, a second stack was added alongside the call stack. On finding a subroutine invocation log, the line number from which the subroutine is called is added to the stack. On finding an exiting statement, the last line number is popped from the stack. By merging these two stacks, we obtain a stack of pairs `(routineName, lineNumber)` indicating which line the API calls should be attributed to in a given routine. Finally, the names of subroutines are stored in a separate dictionary, the keys of which are line numbers. In some specific cases, e.g., a class accessing a static variable of another class, injected constructors are called implicitly. As *Orka* is not yet able to detect such cases, subroutine invocation statements are sometimes missing and there is no information regarding the line number corresponding to the call. However, by comparing the two stack sizes we are able to detect this and a default value is added to the stack to indicate the absence of any information.

IV. Providing Hardware Usage Accounting

In order to account for tail-energy, we use Wi-Fi, a frequently used component. For a given routine, the drain associated to each energy state (active, tail, idle) of the Wi-Fi antenna can be easily computed by multiplying the time spent in this state by the corresponding power drain provided by the power state machine. Therefore, focussing on the time spent in each state rather than the corresponding energy drain would work fine. For each routine, *Orka* should ultimately be able to provide an estimate of the drain caused by Wi-Fi alongside an energy tuple. The main aim is to correlate the Wi-Fi energy activity with routine calls, therefore *Orka* needs to know when routine invocations start and end, and have access to the power drain caused by Wi-Fi at any time. The start-time and end-time of any routine call could be easily obtained by leveraging the `logcat` enter and exit statements and by using an output mode of `logcat` including the timestamps of log messages.

A. Monitoring the power consumption of Wi-Fi

Just like *eprof*, *Orka* needs access to the power consumption of Wi-Fi with the highest sampling period possible. However, to enforce the software-based nature of *Orka*, power-measurement platforms and complex energy models need to be replaced with information provided by the operating system.

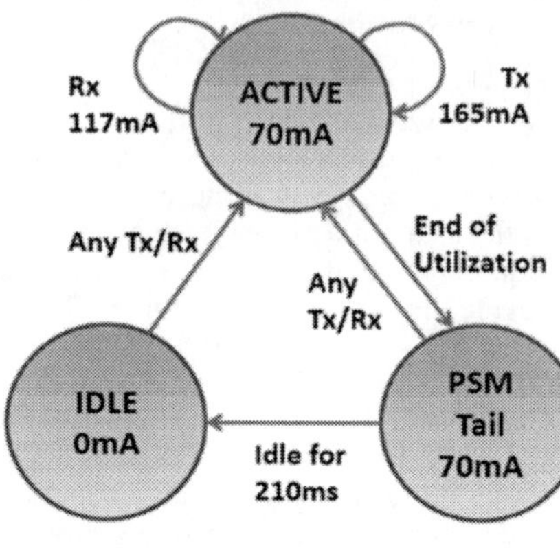

Figure 3. Typical power state machine of a Wi-Fi antenna [19]

As the power behaviour of Wi-Fi antennas can be accurately described by a power state machine, *Orka* could access the power consumption caused by Wi-Fi at any time by monitoring in which energy state the antenna is.

1) Wi-Fi power state machine: The power behaviour of most hardware components, including Wi-Fi, can be accurately described using a power state machine. As shown in Figure 3, Wi-Fi antennas exhibit three power states: active, tail, and idle. By definition, the bandwidth in idle and tail mode, as well as the drain in idle mode are always zero. The bandwidth and drain in active mode, the tail-time and the tail drain differ from one antenna to another. As *Orka* only focusses on energy and not workload, the bandwidth in active mode is not relevant to this work. To this end, the power drain in active mode for a specific device can be obtained using the power profile. From *Android* 5.0, devices come with a power profile providing the energy drain caused by each hardware component in its various energy states. While this includes the drain in active mode for the various components, we found that the power profile doesn't include the drain caused by a component in tail mode and hence is not useful to accurately compute the energy due to tail-behaviour. However, we found in the literature that the drain in tail-mode can be approximated by half the drain in active mode [18] and *Orka* uses this approximation. Finally, we investigated whether we could get the the exact value of the tail-time for any Wi-Fi antenna. Unfortunately, techniques such as the one in [19] involve power-measurement platforms. We hence approximated this value with the one shown in Figure 3. Due to these approximations, *Orka* has access to a complete state machine describing the power behaviour of the Wi-Fi antenna. To monitor the power drain caused by Wi-Fi, *Orka* only needs to know in which power state the antenna is.

2) Monitoring the energy-state: Based on [17], we looked at using the energy tools from the Android Open Source Project. The `batterystats` history contains a timeline of energy related events since the last charge or reset, but it doesn't record switches to the tail state. Hence, we had to use a new approach – monitor events which trigger these switches, namely network traffic events. The Dalvik Debug Monitor Server (DDMS) provides tools to monitor the network traffic in real time for a given application [20]. Using this, we found that network statistics at the application level are stored in `proc/net/xt_qtaguid/stats` [21]. This file contains one line per `(app_uid, tag)` pair, describing the associated network traffic, as shown in Figure 4. A `Python`

```
idx iface uid_tag_int cnt_set rx_bytes
    rx_packets tx_bytes tx_packets
2 wlan0 0 0 200888 1096 79636 888
3 wlan0 0 1 0 0 0 0
```

Figure 4. Extract of `proc/net/xt_qtaguid/stats`

procedure was used to parse this file and return a pair `(rx_bytes, tx_bytes)`, aggregating all the incoming and outgoing traffic induced by a given application. The command `adb shell cat` was used to fetch this file and was able to update the statistics once every 45ms. Network traffic caused by an application with a sampling period of about 50ms could therefore be potentially monitored using an actual device as there is no Wi-Fi emulation on AVDs.

B. *Implementation*

Using these findings, two pieces of software were implemented in order to correlate the routine calls with the Wi-Fi energy activity; (i) a `Python` script to log the switches between the various energy states of the Wi-Fi antenna, and (ii) an analyser to process this data and generate results interpretable by the user. A test application was also written to evaluate the quality of the results generated.

1) Monitoring the Wi-Fi energy state: The pseudo-code in Figure 5 fetches the most recent network data and compares it to a previous one. Any change in this data indicates that network traffic was induced by the application and the antenna switched to active mode. Otherwise, no traffic occurred and if the antenna was in active mode, it switches to tail mode and resets a counter indicating the time when tail mode started. If the antenna was in tail mode, this counter is used to compute the time spent in that mode and switch to idle mode if needed. If the antenna was already in idle mode, it remains so. We ensure that, despite the loss of precision due to the sampling period, the duration of tail mode is never longer than the tail time specified by the transition condition in the state machine.

In order to correlate the routine calls and the energy consumption of the Wi-Fi antenna, *Orka* needs to merge-sort these two files and replay the resulting log file. To accurately sort the timestamps, the clocks used in both logs need to be synchronised. Many networking protocols able to synchronise clocks are available, but most of them are quite complex to use. As *Orka* has to deal with the Android clock, which is used in the `logcat` dump, this clock should also be used to timestamp the switches between the energy states. To this end, the host machine opens the connection with the device and send two commands: the first one to get the epoch time and the second to fetch the file containing the network statistics.

The logs generated by the `networkMonitor` module (Figure 6) will be referred to as `netstats` logs or traces. Comparing these results with those produced by DDMS, we found that the antenna was successfully logged in active or tail mode when traffic was reported by DDMS. Moreover, the sampling period of *Orka* is twice as small as the one of DDMS, as the smallest sampling period offered by this tool is 100ms.

2) Analysing the traffic and execution traces: Once the simulation terminates, *Orka* needs to process this new data to compute the estimates modelling the tail-behaviour induced by each injected routine. To achieve this, the `routine` class is extended to include a dictionary, which stores the time spent by the Wi-Fi antenna in each energy state while the routine was executed. For e.g., a routine executed for 10s without using Wi-Fi will be modelled as: `{'ACTIVE': 0.0, 'TAIL': 0.0, 'IDLE': 10.0}`. The `logcat` data is used to keep track of the call stack and the `netstats` data to update the energy state of the Wi-Fi antenna. The main function parses these two traces and performs a merge sort so as to process the logs chronologically. For each new entry, the time elapsed in the current state since the last entry is added to all the methods in the call stack and the Wi-Fi state and the call stack are then appropriately updated. Once the logs have been fully processed, the results, which include the tuples storing the time spent in each energy state for all the injected routines, are written to disk. Figure 7 shows an example extract.

Input: Active ADB connection to an actual device
Output: Logs of the energy states of the Wi-Fi antenna

Get first network statistics S_0 at current time t_0;
$state \leftarrow$ IDLE;
while *True* **do**
  Get network statistics S_1 at current time t_1;
  if $S_0 \neq S_1$ **then**
    $state \leftarrow$ ACTIVE;
  else if *state* = *ACTIVE* **then**
    $state \leftarrow$ TAIL;
    $tail_{start} \leftarrow t_0$;
  end
  if $state = TAIL$ **and** $t_1 - tail_{start} \geq tail_{time}$ **then**
    Log $(t_0, state)$;
    $t_0 \leftarrow tail_{start} + tail_{time}$;
    $state \leftarrow$ IDLE;
  end
  Log $(t_0, state)$;
  $t_0 \leftarrow t_1$;
  $S_0 \leftarrow S_1$;
end

Figure 5. Network monitor

```
1503657227.287407 ACTIVE
1503657227.323574 TAIL
1503657227.359771 TAIL
1503657227.394717 TAIL
1503657227.428777 ACTIVE
1503657227.467203 TAIL
1503657227.518874 ACTIVE
1503657227.589833 ACTIVE
1503657227.644556 TAIL
1503657227.681700 TAIL
1503657227.716493 TAIL
1503657227.753575 TAIL
1503657227.789498 TAIL
1503657227.829724 TAIL
1503657227.864556 IDLE
1503657227.882524 IDLE
```

Figure 6. Extract of a typical `netstats` output

Building on these new modules, *Orka* is now able to correlate the network activity with the routine calls and attribute the energy usage to them, while taking tail-energy into account.

V. Evaluation

In order to get an insight of the accuracy of the results generated by the implementation, the decision was made to test it against an application designed specifically for this purpose.

1) High-level expectations: As outlined by [7], the power behaviour of the Wi-Fi mostly depends on the total network traffic and on the density of this traffic. In the case when the traffic is particularly dense, the energy drain should be roughly proportional to the total traffic since the antenna will remain

```
MainActivity.<init> {'ACTIVE': 0.0, 'IDLE':
    0.011000156402587 89, 'TAIL': 0.0}
MainActivity$1.onCheckedChanged{'ACTIVE': 0.0,
    'IDLE': 0.03099989891052246, 'TAIL': 0.0}
MainActivity$2.onClick {'ACTIVE': 0.0, 'IDLE':
    0.01699995994567871, 'TAIL': 0.0}
MainActivity.onCreate {'ACTIVE': 0.0, 'IDLE':
    0.13199996948242188, 'TAIL': 0.0}
MainActivity$SendGet.run {'ACTIVE':
    1.5723857879638672, 'IDLE':
    0.8458666801452637, 'TAIL':
    1.626746654510498}
```

Figure 7. Extract of a typical output of the `networkAnalyser`

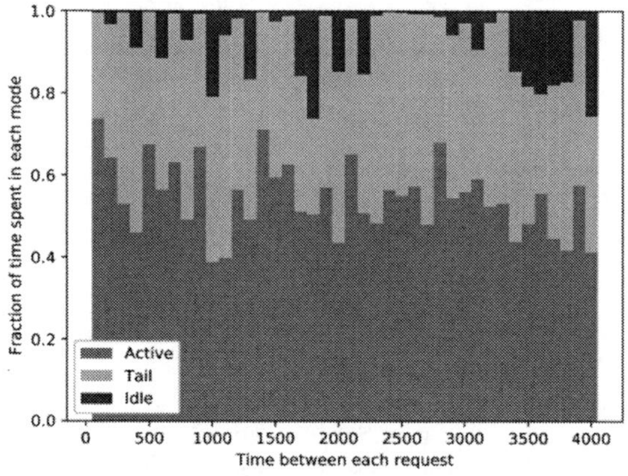

Figure 8. For the method `SendGet.run()`

in active mode and won't exhibit any tail-energy behaviour. However, in the situation when the antenna has to process a small work-flow (sparse traffic) as soon as it enters its idle state, the antenna will spend most of its time in the tail mode. Based on this, we built a test application which allows to generate traffic of various densities and to check whether the results generated by *Orka* are fitting with these principles.

2) Designing a test application: An Android application was created using Android Studio to simply send a `HTTP GET` request every `T` milliseconds to a constant target URL. `T` was initially set to 1000ms and the target URL to `http://www.google.com`. A simple GUI was added later to let the user specify the target URL and the parameter `T`.

3) Running the tests: Orka was then run on this test application using the target URL `http://www.google.com` for various values of `T` between 100ms and 4000ms in steps of 100ms. To this end, `monkeyrunner` scripts were then generated in order to automatically set the right value of `T` and let *Orka* monitor the traffic during 20 seconds. The results generated for $T = 1000ms$ are presented in Figure 7.

4) Analysing the results: At first glance, it was found that only the method `SendGet.run()`, which fires the `HTTP` request was attributed significant network usage, i.e. time in active and tail mode. All other methods were only attributed time in idle mode. This shows that *Orka* was able to detect which method was producing network traffic, and therefore to map the hardware energy usage back to the code. Based on this, it is relevant to compare the energy-state tuples of the method `SendGet.run()` for all values of `T` and to check whether the expectations described in Section V-1 were met.

Figure 8 shows the fraction of the time spent in each mode

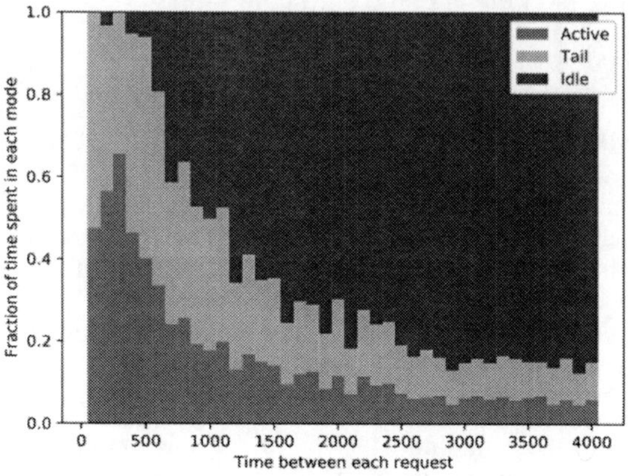

Figure 9. At the application level

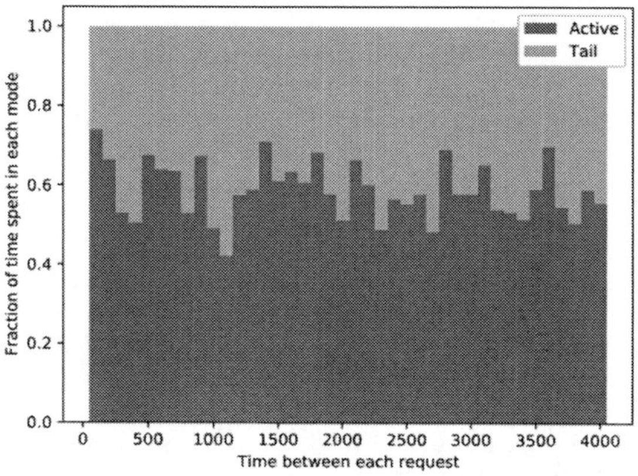

Figure 10. For `SendGet.run()`, ignoring idle mode

depending on the time `T` between each request. It seems that these ratios are constant with respect to the time between each request but include a significant amount of noise. This result may seem surprising at first, as one would expect the fraction of time spent in idle mode to increase with `T`, while the fraction of time spent in tail mode to decrease with `T`. This can be explained by the fact that to compute the energy tuple of a routine, the analyser only focusses on Wi-Fi activity during the execution of this routine. Nevertheless, by definition, tail energy continues beyond the execution of the routine and *Orka* is therefore not accounting for part of the tail energy and of time spent in idle mode. In order to improve the accuracy of the results at the method level, *Orka* needs to implement more complex accounting policies, such as the last-trigger accounting policy, where the routine which last triggered a hardware component will be attributed the energy consumption that follows until another routine accesses this component.

To understand the Wi-Fi activity not attributed to any routine by *Orka*, we looked at the energy tuples at the application level. Figure 9 shows the fraction of the time spent in each mode depending on `T` at the application level. As expected, the fraction of time spent in idle mode increases as the traffic gets less dense but this graph doesn't allow us to draw conclusions about the active and tail modes, although it seems that the fraction of time spent in active and tail modes are similar.

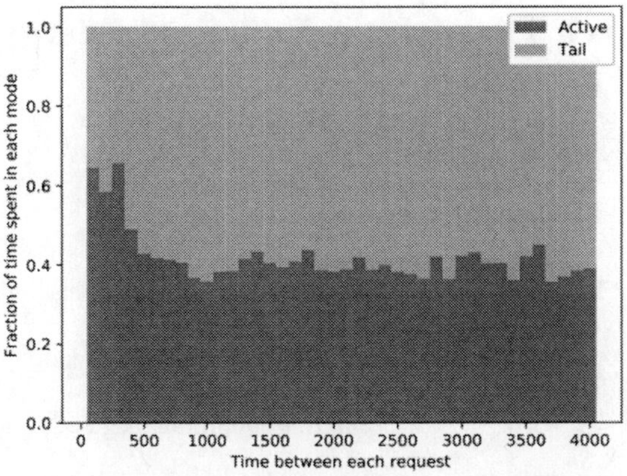

Figure 11. At the application level, ignoring idle mode

Figures 10 and 11 show Figures 8 and 9 redrawn, but with a focus on time spent only in active and tail modes. As we can see, at the method level, as the fraction of time spent in idle mode was small as compared to the time spent in the other modes, Figures 8 and 10 are very similar. However, Figures 9 and 11 show these results at the application level and there *Orka* attributes more time in tail mode as the traffic gets less dense and hence meets the high-level expectations described in Section V-1. Moreover, it seems that the fraction of the time spent in tail mode quickly reaches its maximum of about 60%, for values of `T` higher than 500ms. Finally, according to these results, the Wi-Fi antenna spends at least 40% of the time in tail mode, even for a dense traffic. This would suggest that at least 20% of the drain caused by Wi-Fi is due to tail-behaviour.

VI. CONCLUSIONS AND FUTURE WORK

In this paper, we presented a software-based approach to providing fine-grained energy feedback (at the source-line level) to developers, enabling them to investigate energy bugs effortlessly. This work is also able to map the energy drain caused by the Wi-Fi antenna back to the code and to partially account for the tail-energy. One of the limitations of *Orka* is its heavy reliance on the cost of Android APIs found by [11], which should be updated to have the energy estimate of newer APIs to ensure accurate feedback. Orka operates on the main assumption that the cost of the routines making no calls to the Android API is marginal. However, a routine may not make any calls to the Android API, but instead invoke a subroutine which includes these. To this end, the injector should build a call graph of the user-defined routines, the leaves and nodes of which would be respectively the API calls and the routines. Moreover, the energy estimates generated by monitoring the energy-activity of the hardware should be included in the energy estimates provided by *Orka* at the method-level. Finally, this paper focussed exclusively on Wi-Fi and should include all other hardware components.

REFERENCES

[1] M. V. Heikkinen, J. K. Nurminen, T. Smura, and H. Hämmäinen, "Energy efficiency of mobile handsets: Measuring user attitudes and behavior," Telematics and Informatics, vol. 29, no. 4, 2012, pp. 387–399.

[2] R. Jabbarvand, A. Sadeghi, J. Garcia, S. Malek, and P. Ammann, "Ecodroid: An approach for energy-based ranking of android apps," in Proceedings of the 4th International Workshop on Green and Sustainable Software, 2015, pp. 8–14.

[3] C. Bunse, H. Höpfner, S. Roychoudhury, and E. Mansour, "Choosing the "best" sorting algorithm for optimal energy consumption." in ICSOFT (2), 2009, pp. 199–206.

[4] C. Sahin et al., "Initial explorations on design pattern energy usage," in Proceedings of the 1^{St} International Workshop on Green and Sustainable Software (GREENS), 2012, pp. 55–61.

[5] B. Westfield and A. Gopalan, "Orka: A New Technique to Profile the Energy Usage of Android Applications," in Proceedings of the 5th International Conference on Smart Cities and Green ICT System, SMARTGREENS, 2016, pp. 213–224.

[6] "Orka source code," https://github.com/acornet/orka, retrieved: May 6th, 2018.

[7] A. Pathak, Y. C. Hu, and M. Zhang, "Where is the energy spent inside my app?: fine grained energy accounting on smartphones with eprof," in Proceedings of the 7th ACM european conference on Computer Systems, 2012, pp. 29–42.

[8] D. Li, S. Hao, W. G. Halfond, and R. Govindan, "Calculating source line level energy information for android applications," in Proceedings of the International Symposium on Software Testing and Analysis, 2013, pp. 78–89.

[9] D. Li and W. G. Halfond, "An investigation into energy-saving programming practices for android smartphone app development," in Proceedings of the 3rd International Workshop on Green and Sustainable Software, 2014, pp. 46–53.

[10] C. Sahin et al., "How does code obfuscation impact energy usage?" Journal of Software: Evolution and Process, 2016.

[11] M. Linares-Vásquez et al., "Mining energy-greedy api usage patterns in android apps: an empirical study," in Proceedings of the 11th Working Conference on Mining Software Repositories, 2014, pp. 2–11.

[12] J. Flinn and M. Satyanarayanan, "Powerscope: A tool for profiling the energy usage of mobile applications," in Proceedings of the 2nd IEEE Workshop on Mobile Computing Systems and Applications, 1999, pp. 2–10.

[13] A. Hindle et al., "Greenminer: A hardware based mining software repositories software energy consumption framework," in Proceedings of the 11th Working Conference on Mining Software Repositories, 2014, pp. 12–21.

[14] S. Hao, D. Li, W. G. Halfond, and R. Govindan, "Estimating android applications' cpu energy usage via bytecode profiling," in Proceedings of the 1^{St} International Workshop on Green and Sustainable Software (GREENS), 2012, pp. 1–7.

[15] L. Zhang et al., "Accurate online power estimation and automatic battery behavior based power model generation for smartphones," in Proceedings of the IEEE/ACM/IFIP International Conference on Hardware/Software Codesign and System Synthesis (CODES+ ISSS), 2010, pp. 105–114.

[16] A. Pathak, Y. C. Hu, M. Zhang, P. Bahl, and Y.-M. Wang, "Fine-grained power modeling for smartphones using system call tracing," in Proceedings of the 6th conference on Computer systems, 2011, pp. 153–168.

[17] D. Di Nucci et al., "Software-based energy profiling of android apps: Simple, efficient and reliable?" in Proceedings of the 24th International Conference on Software Analysis, Evolution and Reengineering (SANER), 2017, 2017, pp. 103–114.

[18] "Optimizing Downloads for Efficient Network Access," https://developer.android.com/training/efficient-downloads/efficient-network-access.html, retrieved: March 31^{St}, 2018.

[19] N. Ding et al., "Characterizing and modeling the impact of wireless signal strength on smartphone battery drain," in Proceedings of the ACM SIGMETRICS/International Conference on Measurement and Modeling of Computer Systems, New York, NY, USA, 2013, pp. 29–40.

[20] "DDMS Network traffic tool," https://developer.android.com/studio/profile/ddms.html#network, retrieved: March 31^{St}, 2018.

[21] "Stackoverflow - How does Android tracks data usage per application?" https://stackoverflow.com/questions/31455533/how-does-android-tracks-data-usage-per-application, retrieved: March 31^{St}, 2017.

An Inclusion of Electric Grid Reliability Research Through the Enhanced Energy Informatics Research Framework

Work-In-Progress Paper

Vivian Sultan
Center of Information Systems and Technology
Claremont Graduate University
Claremont, CA
email: vivian.sultan@cgu.edu

Au Vo
Department of Information Systems
San Francisco State University
San Francisco, CA
email: auvo@sfsu.edu

Brian Hilton
Center of Information Systems and Technology
Claremont Graduate University
Claremont, CA
email: brian.hilton@cgu.edu

***Abstract*— Energy Informatics (EI) is the area of research that addresses the application of technology to resolve complex problems within the energy domain. Goebel et al. have provided an EI research framework that attempts to encompass all aspects of EI research. Due to the rapid improvements in EI, there are uncharted research areas that were not incorporated into the framework. Specifically, we posit that grid reliability is an underrepresented research area and should be incorporated into the framework. The goal of this research in progress is to bring forth grid reliability research and present to the community as a viable, important, and bountiful research domain. In this work-in-progress, we raise the need for grid reliability research, define grid reliability, and provide the Enhanced EI Research Framework.**

Keywords-Energy Informatics; Grid Reliability; EI Research Framework.

I. Introduction

The Energy informatics (EI) research covers the use of information and communication technology to address energy challenges [1]. It is the area of research that addresses the application of technology to resolve complex problems in the energy domain. The Department of Energy's Office of Energy Efficiency & Renewable Energy [2] classified challenges within the energy domain into two types: (1) the Transmission System challenges and (2) the Distribution Systems challenges. Transmission Systems challenges include issues in Grid Operations and Grid Stability while Distribution Systems challenges refer to Power Quality and Protection Coordination issues.

Goebel et al. [3] stated that energy efficiency and the renewable energy supply are the two principal types of research movements currently within the energy domain. Energy efficiency research involves studying individual incentives and behavioral dynamics to influence electricity consumers' usage behavior. This first type drives the evolution of smart energy saving systems. The second type of research, renewable energy supply, seeks to resolve challenges that arise with the integration of renewable sources of energy, such as wind and solar power generation into the electric grid. This in turn drives the advancement of smart grids.

In this work-in-progress research paper, we attempt to define grid reliability and present our initial findings of a revised EI framework. These steps are the preludes for our upcoming publication in regards to elucidate an understudied research area within the EI research area: the grid reliability research.

Due to the rapidly changing nature of energy generation, new developments of the electric power network, the incorporation of distributed energy resources into the grid, circuits and equipment overloads, grid reliability research has been underwhelming. In Goebel et al. [3] EI research framework, grid reliability research was considered an inferior topic. Specifically, he considered reliability as one of the segments under the renewable energy supply research movement. However, due to the shift in challenges within the electric utility industry, we argue that grid reliability research should be classified as a separate new type of research.

The rest of the paper is organized as follows. Section II, we raise awareness of grid reliability by putting forth the research need. In Section III, we attempt to define grid reliability. In Section IV, we describe our systematic literature review method. In Section V, we present the results of the review are described, a new enhanced EI framework is proposed, and finally, Section VI contains the conclusions of the paper.

II. Research Need

Since EI is a relatively young field, there are scattered research that handle different aspects of EI, ranging from efficiency, storage, to societal impact of renewable energy. Two seminal work by Watson et al. [4] and Watson and Brodeau [1], and later built upon by Goebel et al [3] are the most early work that attempted to provide a comprehensive EI framework. Other related works focusing on Green Technology [5][6] are included in the framework, but it seems like a subset of reliability has been omitted.

The lack of grid reliability in the current research frameworks set forth a need analysis of grid reliability as a necessary topic in EI. An omission of grid reliability is not only detrimental to the understanding of the electric grid overall but also an impediment for creating a reliable and sustainable energy storage and consumption system. In addition, the research attempts to provide additional research tools that are otherwise overlooked by previous framework. Such inclusion from the domain and the tool perspectives will help facilitate a more complete research agenda with regards to EI research.

Due to a rise in electricity usage via new technologies, such as electric vehicles, circuits and equipment overloads, a significant number of publications from research organizations, governmental bodies, and utility companies have focused on understanding grid reliability, causes of faults, and analyses of power outages events. The National Academies of Sciences [7] has recently published "Enhancing the Resilience of the Nation's Electricity System" in response to the US Congress's call for an independent assessment to "conduct a national-level comprehensive study on the future resilience and reliability of the nation's electric power transmission and distribution system." In addition, the National Academies of Sciences established a committee to conduct the relevant research. Throughout this report, the committee highlighted all elements of grid reliability and resilience, the risks of the system wide failure that will grow as the structure of the power industry becomes more atomized and complex, and laid out a wide range of actions to improve the resilience of the US power system. Analytics (including machine learning, data mining, and other artificial intelligence-based techniques) will play a very important role in response to the diagnosed attacks on the electric grid, failures, or other impairments due to their capability of generating real-time recommendations [7].

In another exemplar research, Adderly examined Department of Energy's (DOE's) power outage data from 2002-2013 and investigated reliability trends [8]. The research objective was to assess the correlation between utilities' reliability and grid investment projects such as the deployment of smart grid assets. Using the deployment of smart meters as a proxy for grid investments, Adderly concluded that the increase of smart meters correlated strongly with the decrease of the frequency of outages [8]. The author acknowledged that due to the presence of confounding variables, the reduction in power outage couldn't be attributed to any specific smart grid investment project.

There are several studies that attempt to understand grid reliability. Mitnick's report prepared for the Electric Markets Research Foundation is another important resource that explains the reasons for concerns about grid reliability [9]. The author suggests that the incorporation of the distributed energy resources into the grid should be carefully managed to minimize the grid reliability risk. Another relevant study conducted by the Lawrence Berkeley National Lab brought the attention to the fact that reliability data trends might not improve due to the addition of smart grid technology, such as automated outage management systems is reporting service interruptions more accurately [10]. Since the study was based on a sample of reliability data from several utilities, the authors did not attempt to make claims about overall power reliability in the US.

With respect to the power outage causes study domain, the majority of the outages in the US are the result of events that occur at the distribution side of the grid. Infrequent outages are caused by the external factors. There are three main causes for the electrical outages: (1) Hardware and technical failure, (2) environment-related, and (3) human errors. In hardware and technical failure, outages are experienced due to equipment overload, short circuits, brownouts, and blackouts, to name a few [11]–[13]. These failures are often attributed to unmet peak usages, outdated equipment, and malfunction of backup power systems [14].

Interrupted power supply is not deemed as a mere inconvenience any longer. As the duration and spatial extent of electricity system outage increase, costs and inconvenience grow. Critical social services, such as medical care, police and other emergency services, and communications systems are relied upon electricity to function at the minimum. Such failure can bring catastrophic outcome; lives can be lost. Grid Reliability is slated to be a preventative research, and the more we understand the causes, the ready we are to implement redundancy and resilience into the electric grid.

To heed this call, we peruse the literature regarding grid reliability to bring forth the knowledge base of the topic. There are two approaches in reviewing the literature, either through a traditional literature review or through a systematic literature review [16]. A systematic literature review is a particularly influential tool in the hands of researchers, since it allows a scholar to gather and recap all the information about research in a specific field [15]. A systematic literature review has the following advantages over a traditional one. First, systematic literature review contains both breadth and focus in the search. Second, the review is rooted in empirical knowledge and evidence rather than preconceived notion. Third, a systematic literature review is replicable [16].

As a first systematic literature review in grid reliability research, this systematic literature review will focus on the different grid reliability topics and their specific characteristic. We intend to use this article to enrich the existing literature reviews and integrate the most recent articles into the body of knowledge.

III. GRID RELIABILITY DEFINITION

The grid reliability topic area aims to address challenges and remove barriers to integrating high penetrations of distributed energy generation at the transmission and distribution levels [2].The subject includes many unaddressed questions such as what makes the grid reliable? Why might reliability degrade? How do the changes in use of the grid that are underway impose such a risk? What can we do to improve the grid reliability? The DOE has recently released its electric grid reliability study, recommending the research prioritization on the development for grid resiliency, and reliability [17]. Additionally, in the same month, the DOE's Office of Electricity Delivery and Energy Reliability announced an initial investment of nearly $900,000 to address the risk and uncertainty of power systems, enabling academic research in the US [18].

There are several definitions with respect to the reliability of the grid. The North American Electric Reliability Council defined reliability as "the degree to which the performances of the elements of the electric system result in power being delivered to consumers within accepted standards and in the amount desired" [19]. Osborne & Cornelia viewed reliability as "the ability of the power system components to deliver electricity to all points of consumption, in the quantity and with the quality demanded by the customer" [10]. Reliability is measured by outage indices as illustrated by the Institute of Electrical and Electronics Engineers (IEEE) Standard 1366. To facilitate a unified view of grid reliability, we define grid reliability as the ability of the electric grid to deliver electricity to customers without degradation or failure. The argument is that today's power systems cannot accommodate significant variable distributed energy generation, for instance, without failure [20].

IV. Systematic Review Methodology

To facilitate an understanding of current grid reliability research, we conduct a systematic review under the framework offered by Kitchenham et al. [21]. There are three stages in a systematic review: the Planning Stage, the Conducting Stage, and the Reporting Stage.

In the Planning Stage, we define the research questions, select the search strategy and parameters. In the Conducting Stage, we conduct the search via two prominent research databases: IEEE Xplore and Web of Science. We have found 503 papers from 2015-2017. After quality assessment, we have perused 231 relevant papers in our systematic review. We compare the results with the search strategy and select the appropriate research to include in the in-depth analysis. Using this knowledge, we set out to finalize our research by revising the EI Framework. That is the product of the Reporting Stage.

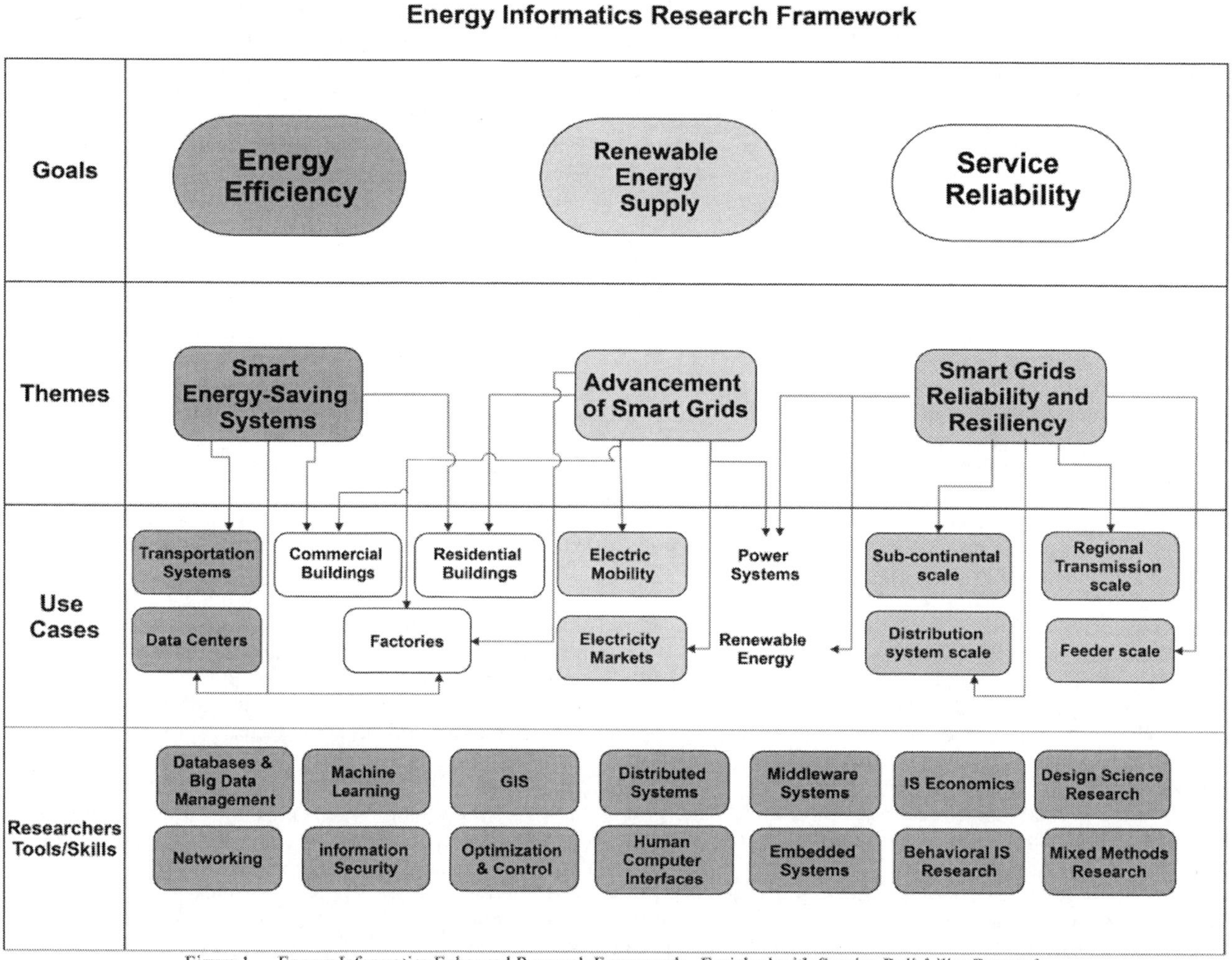

Figure 1. Energy Informatics Enhanced Research Framework - Enriched with Service Reliability Research

V. A PROPOSED REVISED EI FRAMEWORK

We restructured Goebel et al. [3] framework to include service reliability as a third research movement in addition to energy efficiency and the renewable energy supply to recognize this understudied research. In our proposed the Enhanced EI Research Framework in Figure 1, energy efficiency, renewable energy supply, plus service reliability are the three types of research movements within the energy domain. The first theme, Energy Efficiency, drives the evolution of smart energy saving systems, the second theme, Renewable Energy Supply, drives the advancement of smart grids, while the third additional theme, Service Reliability, drives smart grids reliability and resiliency.

In the context of the service reliability research theme, we classify use cases (a collection of possible scenarios) into four fundamental scales: sub-continental, regional transmission, distribution system, and feeder scale. Sub-continental scale can be described as a large, relatively self-contained landmass forming a subdivision of a continent. Within this category, multiple grids, transmission, and distribution systems may exist and be interconnected. Regional transmission is the high-voltage transmission network that enables power to travel long distances from generating units to substations closer to local end-use customers where the voltage is stepped back down and sent into the distribution system for delivery to consumers.

As for the third type of use cases, the electric distribution system moves power from the energy system to the meters of electricity customers. Typically, power is delivered to distribution substations from two or more transmission lines, where it is converted to a lower voltage and sent to customers over distribution feeders. Although distribution system outages tend to be more frequent than those occurring on transmission facilities, the impacts of such outages are smaller in scale and generally easier to repair [7].

Considering the fourth fundamental category of uses cases, they fall under the feeder scale. Customers on radial systems are exposed to interruption when their feeder experiences an outage. In metropolitan areas, these feeders typically have switches that can be reconfigured to support restoration from an outage or regular maintenance. When a component fails in these systems, customers on unaffected sections of the feeder are switched manually or automatically to an adjacent, functioning circuit. However, this still exposes critical services such as hospitals or police stations to potential outages, so these facilities are often connected to a second feeder for redundancy [7].

We also enhanced the Researchers' Tools/Skills category by adding the use of tools such as those of Geographic Information Systems (GIS), Design Science Research, and Mixed Methods research. Analytics through the use of these tools would set out to transform the way we think, act, and use energy. It can help elucidate a root cause of a problem, define a solution through data, and also implement the solution with continuous monitoring and management.

VI. DISCUSSION AND CONCLUSION

In this research in progress, we proposed the Enhanced EI Research Framework to augment the original framework by Goebel et al. [3]. The Revised EI Framework brings forth two important contributions to knowledge. First, our systematic literature review provided a scientific approach and add value to the ever-expanding research in the EI domain. As the old saying goes, prevention is better than cure, we bring forth the reliability to help spark another stream of research focusing on power sustainability and reliable scaling. As renewable sources of energy start to become mainstream, the difficulty is to disperse, manage, and sustain energy storage and consumption in both local and global scale.

Second, we provided tools that were not included in the original framework. The inclusion of GIS as a research tool is imperative to research that focus on location and spatial arrangements of power stations, energy distribution grids and networks, and spatial-enabled events. In addition, with space-time analysis, GIS could assist in understanding the development of power grids through time.

Another important research tool is through the use of Design Science Research. Design Science Research is a type of Information Systems research that focuses on the science of the artifacts [22]–[25]. Researchers utilize two processes: the build and evaluation cycle of Design Science to develop algorithms, methods, and tools to solve real-world problems while relying upon research rigor [22]. Design Science strives when faces with new issues where technology can be the catalyst for change and solution for problems.

As the world has increasingly relied on energy, grid reliability is both an understudied topic and an important one. We hope that with the introduction of the grid reliability research into the Enhanced EI Research Framework, we invite more productive research conversations about this topic. In addition, new tools will help research able to answer existing questions with renewed perspectives, furthering the understanding of EI.

REFERENCES

[1] R. Watson and M.-C. Boudreau, *Energy Informatics*, 1 edition. Green ePress, 2011.

[2] Department of Energy, "Grid Performance and Reliability," *Grid Performance and Reliability*. [Online]. Available: https://energy.gov/eere/solar/grid-performance-and-reliability. [Accessed: 29-Mar-2018].

[3] C. Goebel *et al.*, "Energy Informatics," *Bus. Inf. Syst. Eng.*, vol. 6, no. 1, pp. 25–31, Feb. 2014.

[4] R. T. Watson, M.-C. Boudreau, and A. J. Chen, "Information systems and environmentally sustainable development: energy informatics and new directions for the IS community," *MIS Q.*, pp. 23–38, 2010.

[5] S. Mithas, J. Khuntia, and P. K. Roy, "Green Information Technology, Energy Efficiency, and Profits: Evidence from an Emerging Economy.," in *ICIS*, 2010, p. 11.

[6] V. Dao, I. Langella, and J. Carbo, "From green to sustainability: Information Technology and an integrated sustainability framework," *J. Strateg. Inf. Syst.*, vol. 20, no. 1, pp. 63–79, 2011.

[7] National Academies of Sciences, *Enhancing the Resilience of the Nation's Electricity System*. 2017.

[8] S. Adderly, "Reviewing Power Outage Trends, Electric Reliability Indices and Smart Grid Funding," PhD Thesis, The University of Vermont and State Agricultural College, 2016.

[9] S. Mitnick, "Changing Uses of The Electric Grid: Reliability Challenges and Concerns." Electric Markets Research Foundation, 2015.

[10] J. Osborn and C. Kawann, "Reliability of the US electric system–Recent trends and current issues," *EScholarship - Univ. Calif.*, 2002.

[11] Westar Energy, "What causes power outages – working to improve service reliability." [Online]. Available: https://www.westarenergy.com/outage-causes. [Accessed: 29-Mar-2018].

[12] Diesel Service and Supply, "Causes of Power Failures & Power Outages | Diesel Service." [Online]. Available: http://www.dieselserviceandsupply.com/Causes_of_Power_Failures.aspx. [Accessed: 29-Mar-2018].

[13] Rocky Mountain Power, "Key Causes of Power Outages." [Online]. Available: https://www.rockymountainpower.net/ed/po/or/kcopo.html. [Accessed: 29-Mar-2018].

[14] K. Chayanam, "Analysis of Telecommunications Outages Due to Power Loss," Ohio University, 2005.

[15] G. Spanos and L. Angelis, "The Impact of Information Security Events to the Stock Market," *Comput Secur*, vol. 58, no. C, pp. 216–229, May 2016.

[16] R. Mallett, J. Hagen-Zanker, R. Slater, and M. Duvendack, "The benefits and challenges of using systematic reviews in international development research," *J. Dev. Eff.*, vol. 4, no. 3, pp. 445–455, 2012.

[17] T. Profeta, "The Climate Post: Grid Reliability Study Released As Climate Change Panel Disbands," *Huffington Post*, 24-Aug-2017. .

[18] R. Laezman, "Office of Energy Invests in Grid Reliability Research," *Office of Energy Invests in Grid Reliability Research*, 2017. [Online]. Available: https://www.ecmag.com/section/systems/office-energy-invests-grid-reliability-research. [Accessed: 29-Mar-2018].

[19] E. Hirst and B. Kirby, "Bulk-power Basics: Reliability and Commerce," in *Consulting in Electric-Industry Restructuring*, Oak Ridge, Tennessee, 2000.

[20] Department of Energy, "Chapter 3 — Enabling Modernization of the Electric Power System | Department of Energy." [Online]. Available: https://energy.gov/under-secretary-science-and-energy/downloads/chapter-3-enabling-modernization-electric-power-system. [Accessed: 29-Mar-2018].

[21] B. Kitchenham, *et al.*, "Systematic literature reviews in software engineering–a systematic literature review," *Inf. Softw. Technol.*, vol. 51, no. 1, pp. 7–15, 2009.

[22] A. R. Hevner, S. T. March, J. Park, and S. Ram, "Design science in information systems research," *MIS Q.*, vol. 28, no. 1, pp. 75–105, 2004.

[23] A. Hevner and S. Chatterjee, *Design Research in Information Systems*, vol. 22. Boston, MA: Springer US, 2010.

[24] S. T. March and G. F. Smith, "Design and natural science research on information technology," *Decis. Support Syst.*, vol. 15, no. 4, pp. 251–266, 1995.

[25] J. G. Walls, G. R. Widmeyer, and O. A. El Sawy, "Building an information system design theory for vigilant EIS," *Inf. Syst. Res.*, vol. 3, no. 1, pp. 36–59, 1992.

Electric Substation Emergency Disaster Response Planning Through the Use of Geographic Information Systems

Vivian Sultan
Center of Information Systems and Technology
Claremont Graduate University
Claremont, CA
email: vivian.sultan@cgu.edu

Au Vo
Department of Information Systems
San Francisco State University
San Francisco, CA
email: auvo@sfsu.edu

Brian Hilton
Center of Information Systems and Technology
Claremont Graduate University
Claremont, CA
email: brian.hilton@cgu.edu

Abstract— **There are many reasons behind power failure that we are facing today. When examining some of these reasons, the research team found that an electric substation failure could cause massive power outages. In this research-in-progress paper, we investigate the important relationship between electric substations and the important infrastructure they support. Specifically, we try to answer the research question: "Which important facilities would be affected by the power outage from an electric substation level?" To answer the question, we build a proof of concept Decision Support Tool using Geographic Information Systems to inform emergency responders which important facilities need to be attend to first. We select nursing homes and emergency facilities in Southern California, USA as the facilities of interest for the proof of concept.**

Keywords- *Emergency Disaster Response; GIS; Power outage.*

I. INTRODUCTION

Electrical power, within a short time, has become a necessity of modern life. Our work, healthcare, leisure, economy, and livelihood depend on the constant supply of electrical power. Even a temporary power outage can lead to relative chaos, financial setbacks, and possible loss of life. Our cities dangle on electricity and without the constant supply from the power grid, pandemonium would break loose. Power outages can be especially tragic when it comes to life-support systems in places like hospitals and nursing homes or systems in synchronization facilities such as airports, train stations, and traffic control. Though our lives are interwoven with electricity, outages happen more often that we perceive. An estimated economic cost of power interruptions to U.S. electricity consumers is $79 billion annually in damages and lost economic activity in a 2006 report [1]. A preliminary update of the report put the number approximately at $110 billions in 2014 [2].

As a result, power failure should be treated with the urgency it deserves. In general, there are three causes of power outages: (1) technical failure in the hardware components, (2) environmental-related issues, such as winds, tornadoes, animals, and (3) human-induced errors, e.g., acts of terrorism. The devastating effects of power outage have prompted county-level, state-level, and federal-level investigations [3]–[6]. While maintaining the overall grid reliability is an important issue and being pursued by other researchers, this article investigates a specific issue in power outage, especially in the sudden outage at the substation level.

Power failure at the distributing site such as at a substation is an event that could cause massive power outage. For instance, a 2017 power outage in San Francisco shut down power to businesses, an important rail station, a federal court, and 90,000 inhabitants [7]. The electrical substation is the part of a power system in which the voltage is transformed from high to low or low to high for transmission, distribution, transformation and switching [8]. Substations are normally owned and operated by an electrical utility and are generally unattended, relying on Supervisory Control and Data Acquisition (SCADA) for remote supervision and control. Power transformers, circuit breakers, bus bars, insulators, lighting arrestors are the main components of an electrical substation [8]. Any mechanical failure of these components could lead to a power outage.

In this research-in-progress paper, we investigate the important relationship between electric substations and the crucial infrastructure they support. Specifically, we try to answer this research question: "Which important facilities would be affected by the power outage from an electric substation level?" To answer the question, we built a proof of concept Decision Support Tool using Geographic Information Systems (GIS) to inform emergency responders which facilities need to be attend to first. We select nursing homes and emergency facilities in Southern California, USA as the facilities of interest for the proof of concept.

In Section II, we examine utilities emergency response planning and GIS applications in various sectors to establish GIS as an appropriate solution in the research endeavor. In Section III, we detail steps in which we design a GIS model design to build the proof of concept and present it. In Section IV, we discuss our findings.

II. BACKGROUND LITERATURE REVIEW

In this section, we discuss about the need for a GIS application in dealing with the emergency response planning in the electric utility domain. First, we describe the current emergency response planning landscape in the domain, then we briefly introduce GIS and demonstrate the utility of GIS in different sectors. With the proven usefulness in planning and responding to disasters, the electric utility domain could also be benefited from the GIS technology.

A. Electric Utilities Emergency Response Planning

Electricity is a crucial product many of us take for granted. We scarcely think about it, unless we don't have it. Because electricity plays such an important role in our lives, we rely on electric companies to provide a reliable supply of on-demand power. Companies constantly plan for emergency situations that could impact their ability to generate or deliver power. Overall, the industry has a strong track record of maintaining high levels of reliability [9].

No matter how well the industry is prepared, incidents of power outage still happen. As a result, every electric company has a contingency plan for dealing with power outages. In this plan, there is a detailed instruction for restoring electricity. Typically, one of the first steps a company takes – to respond to significant outages – is to identify where the impacted areas are and who are in danger. Restoration then proceeds based on established priorities. If business operations or households are disrupted, customers expect to know how long they will be impacted. Then, estimated restoration time will be established, monitored, adjusted, and communicated to impacted customers. Regulators and local government officials will also be notified regarding the outage and the impact. All in all, electric utilities strive to meet customer needs through effective risk assessment, mitigation, preparedness, response and communications.

Despite the fact that the electric utilities are doing their parts in restoring service, communities also partake in the emergency response to ensure safety for the community members. However, the practice of emergency response planning varies considerably among communities. In some, the planning process is quite formal: there is a specific assignment of responsibility to an office having an identifiable budget. In other, it is informal: responsibility is poorly defined and a limited budget is dispersed among many agencies [10]. Therefore, there is a need to understand the impact of power outage in the community and which important facilities are immediate affected, especially in the communities where informal emergency response is the norm. With a more holistic view of an emergency event and its related geographical areas, facilities, and population, emergency responders can act more effectively.

B. Geographic Information Systems

The Environmental Systems Research Institute defines GIS as a class of tools for seizing, storing, analyzing, and demonstrating data in relation to its position on the Earth's surface [11]. Analysts utilize GIS to view different objects' locations and study their relationships. Satellite as well as tabular data can be entered into GIS for a single map display. GIS applications include recognizing site locations, mapping topographies and also developing analytical models to forecast events [11].

The utility companies have started to employ GIS but primarily as a display tool and not so much as an analytical tool [12]. There are several applications that exist demonstrating the usefulness of GIS, especially in California: Pacific Gas and Electric Company's Solar Photovoltaic and Renewable Auction Mechanism Program Map, Southern California Edison's Distributed Energy Resource Interconnection Map, and San Diego Gas & Electric's Interconnection Information Map [13]. With the advent of spatial analysis and the importance of geographic nature of emergency response, GIS has started to gain more popularity, thanks to the utility of the tools in solving pressing problems in this field.

C. GIS Applications In Various Sectors

GIS assists with public works departments' budgets and maintenance work prioritization [14]. Michael Isun, an engineer technician for Public Works Road Operations Division, stated that "We use GIS throughout our entire operation, from data collection and asset management to maintaining our predictive models and developing our annual budgets" [14].

In addition to the application of GIS in public works, it has been employed as a forecasting tool to stop and control the spread of infections and diseases in public health. Idowu et al. [15] discussed an instance where GIS was employed to predict the malaria epidemic hot-spots in Nigeria. In this scenario, GIS succeeded to stop the malaria epidemic in a country where malaria incidents were excessive with 8-12.5% mortality rate [15].

Considering GIS success in solving significant problems for public work and public health, we chose GIS to create our Decision Support proof of concept. Not only can GIS be used as a display tool, but it also can be used in analytics and modeling. The proof of concept in Section III will highlight how GIS could aid emergency responders in a case of substation power failure.

III. PROOF OF CONCEPT DESIGN

In this section, we delineate our steps in collecting data and designing a prototype for the GIS application. First, we describe the data that we used and how we got them. Second, we state the software that we utilized to create the proof of concept. Lastly, we display the design steps we took to create the prototype.

A. Data Collection

For the design of the proof of concept, we limit our data collection to the spatial extent of an area that encompasses five cities reside in the San Gabriel Valley area of Los Angeles County, Southern California: Claremont, La Verne, Glendora, San Dimas, and Pomona. We obtain the nursing home data from the Homeland Infrastructure Foundation. There are three types of nursing homes: continued care, retirement community, and assisted living. Given our spatial extent, we chose to include all three types. In addition, we extracted the Urgent Care Facilities from the same dataset. We excluded hospitals in our dataset, only focusing on medical clinics. In addition, we relied upon the US Census Tract Parcel data to provide us with the commercial and residential areas of the spatial extend. While there is an interest of expanding study into the residential demographics, it is outside the scope of our study.

For the substation data, we utilized the Southern California Edison (SCE)'s Distributed Energy Resources interconnection Map (DERiM) Capacity Analysis Data. Since Southern California is serviced by one public company, SCE, this map is adequate to understand the overall electricity makeup of the region. The SCE's DERiM includes a majority of the electricity components: power electric lines, the capacity analysis in kilowatts by circuit line, and the substation locations. The data was utilized to situate and develop an area that is proximal with electric power distribution centers and substations to measure which specific areas would be affected by a potential power outage or outage caused by a natural disaster.

B. Design Software

The project utilized ArcGIS mapping platform for windows desktop. ArcGIS is used for creating and using maps, compiling and editing geographic data, analyzing mapped information, sharing and discovering geographic information using maps and geographic information in a range of applications, and managing geographic information in a database. We used the ArcGIS 10.3 developed by Environmental Systems Research Institute (ESRI) for designing the proof of concept.

C. Design Steps

First, we create a map with the spatial extent that encompasses five cities reside in the San Gabriel Valley area of Los Angeles County, Southern California. We then locate all nursing homes, urgent care facilities, power substations, and power lines in the vicinity, the perform spatial joins to create linkages between them. We then proceeded to color-coding to separate each power substations and its connected facilities. Figure 1 is the final result.

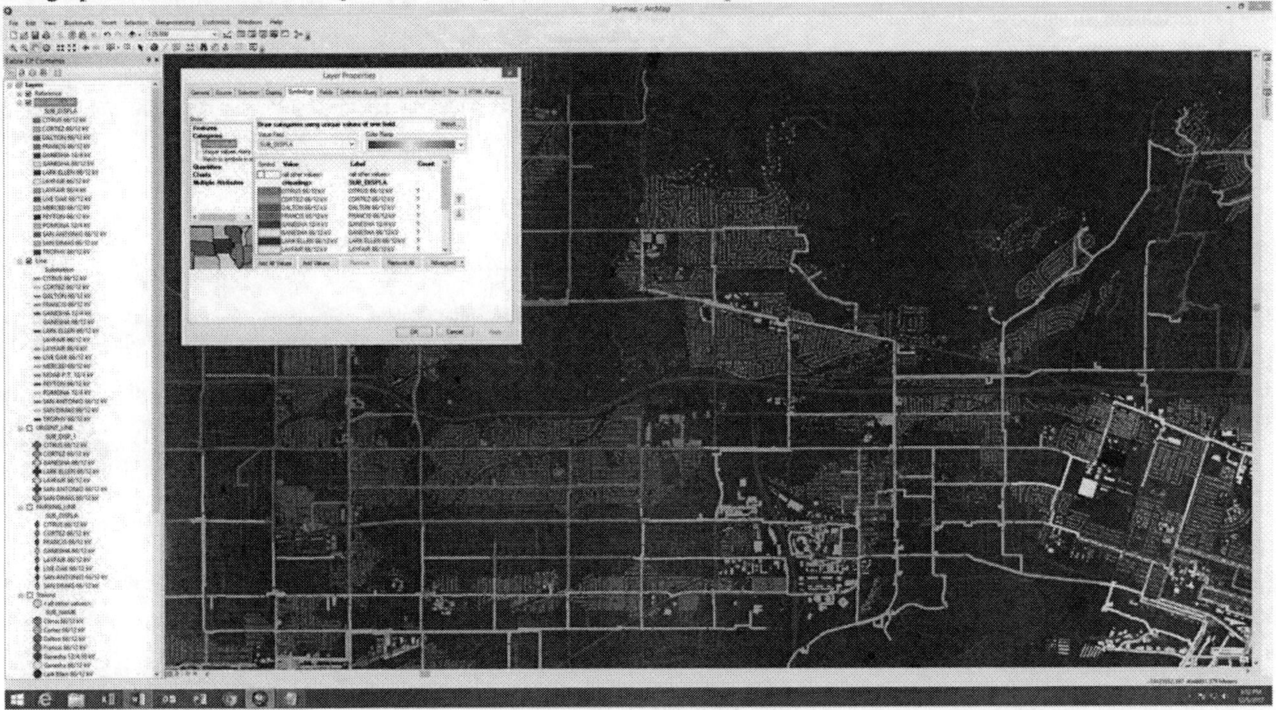

Figure 1. Design Model Output

IV. Discussion and Conclusion

In this research, we build a proof of concept to create linkages between power stations, power lines, nursing homes, and urgent care facilities. The outputs of this study show each substation and its connected lines with nursing homes and urgent care facilities through different color-coding. This could become a tool in which inter-city governments can use to coordinate to response in an emergency to power failure at the substation level. In other words, the prototype provides locations of important facilities that are affected by the power outage. The coordination effort could be diverted to assist the most pressing facilities, as demonstrated by the use of nursing homes and urgent cares in the prototype.

In future iterations of the prototype, we envision a full-scale emergency response platform for different types of critical facilities targeting the most vulnerable populations, such as children hospitals and hospice. The platform can also assist different non-critical facilities but important, e.g., schools, administrative governmental buildings, and churches. In addition, this emergency response could provide utility companies with the understanding in regards to prioritization of a power backup based on the specific needs of each community.

REFERENCES

[1] K. H. LaCommare and J. H. Eto, "Cost of Power Interruptions to Electricity Consumers in the United States," *Energy Int. J.*, 2006, 1845-1855.

[2] J. Eto, "The National Cost of Power Interruptions to Electricity Consumers - An Early Peek at LBNL's 2016 Updated Estimate," in *IEE PES*, Boston, MA, 2016.

[3] J. M. Lee Morgan, "Blackout sparks multiple investigations," *sandiegouniontribune.com*, Sep-2011. [Online]. Available: http://www.sandiegouniontribune.com/news/watchdog/sdut-power-surges-back-on-2011sep09-story.html. [Accessed: 26-Mar-2018].

[4] Governor Cuomo Press Office, "Governor Cuomo, Displeased with Length of Power Outage, Directs Investigation into Rochester Gas & Electric's Preparation For and Response to Windstorm," *Governor Andrew M. Cuomo*, 11-Mar-2017. [Online]. Available: https://www.governor.ny.gov/news/governor-cuomo-displeased-length-power-outage-directs-investigation-rochester-gas-electrics. [Accessed: 26-Mar-2018].

[5] A. Morris, "Senator Ricardo Lara Calls for Independent Investigation of Long Beach Power Outages," *Long Beach Post*, 25-Jul-2015.

[6] Y. Steinbuch, "FBI joins probe into power outage at Atlanta airport," *New York Post*, 19-Dec-2017.

[7] N. Chandler, "Substation Failure Causes Massive Power Outage in San Francisco," *Transmission & Distribution World*, 26-Apr-2017. [Online]. Available: http://www.tdworld.com/substations/substation-failure-causes-massive-power-outage-san-francisco. [Accessed: 26-Mar-2018].

[8] US Department of Agriculture, "Design Guide for Rural Substations." United Stated Department of Agriculture, Jun-2001.

[9] National Academies of Sciences, *Enhancing the Resilience of the Nation's Electricity System*. 2017.

[10] US Department of Homeland Security, "National Response Framework." US Department of Homeland Security, May-2013.

[11] ESRI, "Lesson 1: Why GIS?," in *Understanding GIS--The Arc/Info method*, 6th ed., Redlands, CA: Environmental Systems Research Institute, 1992.

[12] V. Sultan, "Exploring geographic information systems to mitigate America's electric grid traffic congestion problem," in *Computational and Business Intelligence (ISCBI), 2016 4th International Symposium on*, 2016, pp. 74–79.

[13] California Public Utilities Commission, "Distribution Resources Plan (R.14-08-013)," *Distribution Resources Plan*, Jul-2015. [Online]. Available: http://www.cpuc.ca.gov/General.aspx?id=5071. [Accessed: 26-Mar-2018].

[14] D. Totman, "Model predictions: GIS helps public works manage assets," *American City and County*, 17-Apr-2013. [Online]. Available: http://americancityandcounty.com/gis-amp-gps/model-predictions-gis-helps-public-works-manage-assets. [Accessed: 26-Mar-2018].

[15] A. P. Idowu, N. Okoronkwo, and R. E. Adagunodo, "Spatial predictive model for malaria in Nigeria," *J. Health Inform. Dev. Ctries.*, vol. 3, no. 2, 2009, 30-36.

Battery Storage Integration Into the Electric Grid

Vivian Sultan, Ahmed Alzahrani, Hind Bitar, and Brian Hilton
Claremont Graduate University CISAT
Claremont, USA
E-mail: Vivian.Sultan@cgu.edu

***Abstract* - Energy storage is expected to flourish considerably, requiring large investments in tools and instruments that support its integration into the electric grid. This paper addresses the research question: "How can we determine the optimal locations for energy storage for proper integration into the electricity system?" Therefore, we have developed a conceptual framework for decision-making that caters to utilities for it considers the impacts of the energy storage placement on the grid, and the electric circuit capacity constraints. Additionally, we have built a prototype of Geographic Decision Support Systems (GDSS) to aid in the battery storage systems location choices and provide actionable information for utilities and other stakeholders who are concerned about the grid reliability and the urgency of energy storage deployment as a distributed energy resource.**

Keywords- battery storage; electric grid; circuit capacity; optimal location; conceptual framework.

I. INTRODUCTION

Electric utilities in the United States face serious challenges, which have raised concerns about the uncertain future for America's electric grid. The electricity demand is rising while the existing electric infrastructure is deteriorating. Environmental trends force regulators to institute new carbon restraints, and the penetration of distributed energy continues to grow exponentially. To employ the smart grid and resolve these challenges, utilities will need to integrate additional components into the grid. Smart meters, outage management systems, distribution managements systems, sensors, data management systems, and energy storage systems are all examples of the additional components to be incorporated into the grid [1].

Grid reliability is the greatest concern resulting from challenges facing electric utilities. According to the North American Electric Reliability Council (NERC), reliability is "the degree to which the performances of the elements of the electric system result in power being delivered to consumers within accepted standards and in the amount desired" [2]. Osborne and Cornelia [3] view reliability as "the ability of the power system components to deliver electricity to all points of consumption, in the quantity and with the quality demanded by the customer" [3]. Reliability is measured by outage indices as illustrated by the Institute of Electrical and Electronics Engineers (IEEE) Standard 1366. The argument is that today's power systems cannot accommodate significant variable distributed energy generation, for instance, without failure [4].

According to the United States Department of Energy, energy storage technology can help contribute to the overall system reliability as wind, solar, and other renewable energy sources continue to be added to the grid. Additionally, storage technology will be an effective tool in managing grid reliability and resiliency by regulating generation fluctuation and improving the grid's functionality. Storage can provide redundancy options in areas with limited transmission capacity, transmission disruptions, or volatile demand and supply profiles [4]. In addition, the storage market might foster a strong manufacturing base of high-tech electric energy storage devices in the United States, and this capability can enhance America's export opportunities [4]. Further, storage will help promote America's energy independence and reduce carbon emissions by enabling more efficient adoption of renewable energy sources.

An August 2013 White House report, written in conjunction with the Office of Electricity Delivery and Energy Reliability, detailed the vital role that energy storage would play in improving grid resiliency and robustness related to weather outages and other potential disruptions [5]. Energy storage is expected to flourish considerably, requiring large investments in tools and instruments that support its integration into the electric grid. Different States have started new policy incentives and processes to address the integration need. The state of California, for example, has adopted a Roadmap which focuses on addressing three main challenges associated with energy storage adoption: "1) Expanding revenue opportunities 2) Reducing costs of integrating and connecting to the grid 3) Streamlining and spelling out policies and processes to increase certainty" [6]. As a result, Energy storage will be beneficial to all parts of the electricity system as shown in Figure 1 [7].

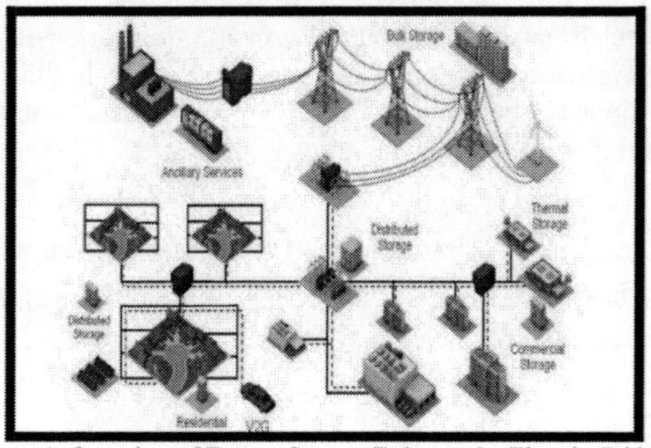

Figure 1. Overview of Energy Storage Roles on the Electric Grid [7]

In this paper, we discuss energy storage systems as previous literature showed insufficient attempts by researchers to provide solutions that can assist in the undertaking of integrating storage systems into the grid. Considering that energy storage is a critical component to be added to the power network and the urgency of energy storage deployment, this paper addresses the research question: "How can we determine the optimal locations for energy storage for proper integration into the electricity system?"

A Geographic Information System (GIS) provides the tools and integration ability to support the smart grid. GIS can used to highlight optimal locations for different components within the network. With the rollout of a smart grid, utilities need "to determine the right location for sensors, communication-marshaling cabinets, and a host of other devices" [1]. GIS provides the proper instruments to perform these design services, especially as the optimal locations depend greatly on the existing infrastructure. Considering the GIS current role in managing the electric grid and the previous research, we posit GIS will play a strong role in the placement of smart grid components such as battery storage systems. GIS can certainly contribute to the transformation of the grid "from a largely passive and blind system to an interactive, intelligent one" [1].

Hence, the objective of this research is to develop a prototype of GIS-based Decision Support System solution, which is an elegant, interesting, and novel solution to assist with the placement of battery storage systems by finding the optimal locations considering the electric grid constraints, the deployment requirements and the potential benefits to the grid. The paper is based on the process steps in Takeda, et. al.'s design cycle to create an artifact/solution [8]. The authors used the three design science research cycles of relevance, design, and rigor [9] to perform each of the Takeda, et. al., process steps leading to the final framework and prototype in this paper.

Takeda et al. [8] cycle has five main steps/phases, which are the awareness of the problem, suggestion, development, evaluation, and conclusion [8]. This paper is organized based on the five phases. In introduction and literature review section, we discussed the awareness of the problem. In the conceptual framework and prototype sections, we covered the suggestion phase by explaining the decisions that we have made to develop the artifacts. In the research approach section, we have indicated the steps taken to develop and create the artifacts as outlined in the development phase. In the evaluation section of the paper, we evaluated the artifacts. In the last section, we concluded our research and offered future research directions.

II. Background Literature Review

This section discusses the recent researches that had been conducted in the field.

A. Energy Storage Technologies and Categories

Energy storage can be defined as the capture of energy produced at one time for use at a later time [10]. It is required to level peak electricity generation. We can divide energy storage technologies into four categories based on discharge times [11]. These categories are useful in conceptualizing the role of storage on the grid. According to the International Electro technical Commission [11], either energy storage technologies can discharge enormous amounts of power, but only for a short time (energy storage with the highest power densities and lower energy densities), or it can discharge energy for a long time, but cannot provide massive amounts of power immediately (technologies with the highest energy densities and lower power densities) [11]. We summarized battery storage technology types in Table I based on their discharge time, use case, and energy-to-power ratio. Additional information about energy storage types can be found in Table A1 in Appendix 1 along with their definitions and examples.

TABLE I. ENREGY STORAGE TECHNOLOGIES [11]

Energy Storage Resources	Use	Discharge Time	Energy-to-Power Ration (kWh/kW)	Examples
Short Discharge Time	Provide instantaneous frequency regulation services to the grid	Second or minutes	Less than 1	Double layer capacitors (DLCs), superconducting magnetic energy storage (SMES), and flywheels (FES)
Medium Discharge Time	Useful for power quality and reliability, power balancing and load following, reserves, consumer-side time-shifting, and generation-side output smoothing. May be designed so as to optimize for power density or	Minutes to hours	Between I and 10	Lead acid (LA), lithium ion (Li-ion), and sodium sulphur (NaS), flywheels may also be used

	energy density.			
Medium-to-Long Discharge Time	Useful primarily for load-following and time-shifting, and can assist RE integration by hedging against weather uncertainties and solving daily mismatch of RE generation and peak loads.	Hours to days	Between 5 to 30	Pumped hydro storage (PHS), compressed air energy storage (CAES), and redox flow batteries (RFBs) which are particularly flexible in their design
Long Discharge Time	Useful for seasonal time shifting (storing excess generation in the summer and converting it back to electricity in the winter).	Days to months	Over 10	Hydrogen and synthetic natural gas (SNG)

B. Types and growth of battery storage market

In the United States, Lithium-ion batteries dominate the energy storage market, accounting for more than 90% of the market share. California, Delaware, Illinois, Indiana, Kentucky, Maryland, Michigan, North Carolina, Ohio, Pennsylvania, Tennessee, Virginia, West Virginia and the District of Columbia together account for about 81% of U.S. energy storage deployment in MW terms. Researchers expect the U.S. energy storage market in year 2021 to grow by almost eight times its 2016 market size [12]. Because of the market growth, there are many vendors competing in various segments. Figure 2 shows the latest taxonomy of energy storage vendors where Tesla and few other companies play a significant role in the U.S. market.

Figure 2. Energy Storage Vendor Taxonomy [12]

Battery storage can be classified into residential, nonresidential, and utility-scale storage types. There are different energy storage system vendors for different storage types. In this paper, we only focus on the medium and the medium-to-long discharge time technologies in the Utility scale segment. Utility-scale storage has been viewed as the key technology, which will help to bring the nation's utility grid into the 21st century. It will play a significant role in meeting the challenges facing electric utilities by improving the operating capabilities of the grid, lowering cost and ensuring high reliability, as well as deferring and reducing infrastructure investments. Utility-scale storage can be also instrumental for emergency preparedness because of its ability to provide backup power, as well as grid stabilization services [13].

C. Optimal locations

Identifying optimal locations for energy storage is a problem that utility companies currently face. Several studies have tried to solve this problem using different methods. For example, Bose, Gayme, Topcu, and Chandy [14] conducted a study to optimally place large-scale energy storage units in power grids considering both conventional and wind generation [14]. The storage type and size played an important part in determining the optimal distributed storage location. The authors developed and tested an algorithm considering different portfolios of wind and conventional generation resources. The results showed that "changes in optimal storage siting configurations remain reasonably consistent regardless of the generation mix and these results are robust to changes in total storage capacity and transmission line limits" [14].

Other scholars have also discussed the optimal location issue associated with the renewable energy sources. Another example was a study that conducted by Meneses de Quevedo, and Contreras [15]. According to the authors, the renewable energy sources are growing rapidly, which causes a problem in the network operation due to higher network constraints [15]. To solve the problem, an Energy Storage System (ESS) is considered as a solution. The authors in this study discussed the uncertainty associated with the nature of wind, network load, and price of ESS in optimizing distribution system costs. The researchers used a Mixed Integer Linear Programming

(MILP) approach as a method to identify the best locations for storage [15].

Additionally, Fernández-Blanco et al. [16] discussed the problem of energy storage (ES) optimal locations in large transmission grids considering two perspectives: (1) a System Operator and (2) a profit-seeking ES owner [16]. A System Operator has the expertise to determine the locations and sizes of energy storage to minimize the investment and the operational costs. The researchers mainly focused on cost minimization, so they combined "the system costs of the [System Operator] and the profits collected by the ES owner into a single objective function subject to the technical constraints" [16]. The researchers solved the problem by applying MILP. Then, they reengineered the problem into a multi-objective optimization with the perspective from both the system operator and the storage owner [16].

Energy storage can also reduce the variability, shift load time, and control ramping time. Xu [17] discussed and provided a solution to determine the optimal location of energy storage using the simulation tool MATPOWER [17]. This study developed a general procedure to optimize the location with the use of point estimation method in order to have more credible results.

GIS is a catalyst for improving multiple facets of smart grids. For instance, Resch et al. have integrated GIS-based modeling into the energy system to address the renewable energy infrastructure planning [18]. Sultan et al. [19] used GIS to optimize the locations of a distributed energy resource such as solar panels [19]. Similarly, Sultan et al. [20] investigated the power grid reliability incidents/power outages and their correlation with the infrastructure age by using GIS-based modeling [20]. Therefore, GIS enhances the research inquiries in the smart grid domain.

Nazaripouya et al. [21] urged researchers to consider tackling the nascent energy storage optimal location research area [21]. With the versatility of GIS, incorporating GIS to determine optimal locations of energy storage in the grid is essential for enhancing the knowledge base and provide a novel solution. Therefore, our paper employs a GIS-approach as the key research vehicle.

III. Theory And Research Approach

The authors followed and used Hevner et al. [9] Design Science Research (DSR) framework to develop the conceptual framework for placement of utility scale energy storage. According to Hevner et al. [9], there are three important cycles in the DSR, which are relevance, design, and rigor cycles, as it shown in Figure 3 [9]. The authors used these cycles to develop the framework by searching the background literature. Moreover, applying the location theory, developing and evaluating the artifacts were the further steps taken by the researchers to determine the optimal locations for placement of utility scale energy storage.

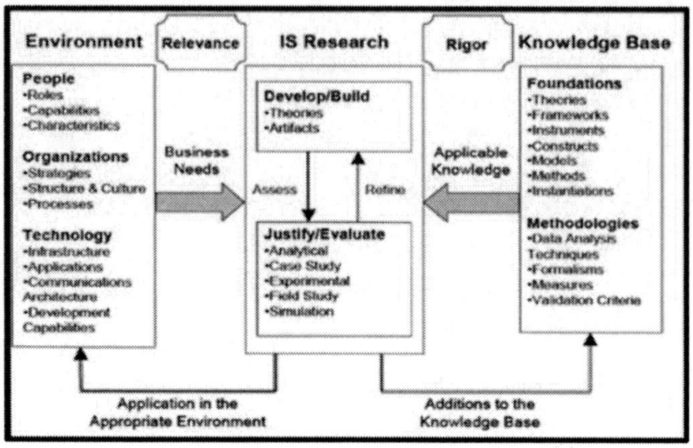

Figure 3. Design Science Research (DSR) Framework in Information Systems Research [8]

Location theory is used to address questions of what economic activities such as energy storage should be located where and why [22]. This theory is based on microeconomics, which studies the behavior of individual and firms in making decisions concerning the allocation of limited resources. Firms, such as utility companies should choose locations that raise their profits and advance the benefits to the grid. Therefore, as is shown in Figure 4, we have chosen the location theory since it considers the firms' perspective to answer what and where question, which matches our research goal. We have classified the site selection factors and structured the process of the energy storage site selection. Location theory's key objective is to explain why specific economic activities prefer to establish themselves in particular areas and locations by allocates with what (energy storage) and where (area, location, or regions).

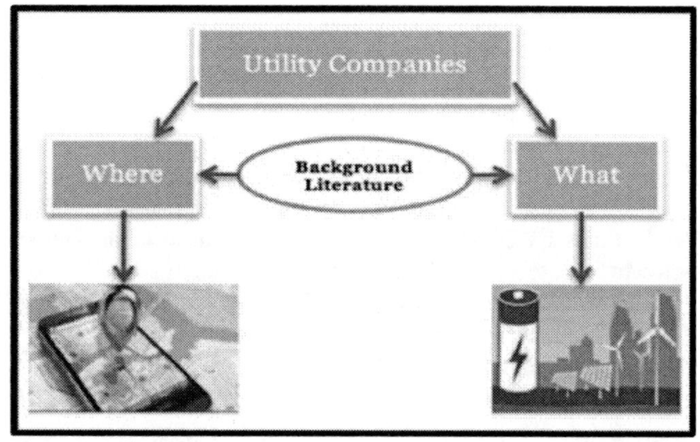

Figure 4. Utility Scale Energy Storage Location Decision Guided by Location Theory

A. *Artifact 1: Conceptual Framework*

The researchers used and studied the background literature as a research method to develop the conceptual framework for decision-making that caters to utilities, for it considers the impacts of the energy storage placement on the grid, and the electric circuit capacity constraints as is shown in Figure 5.

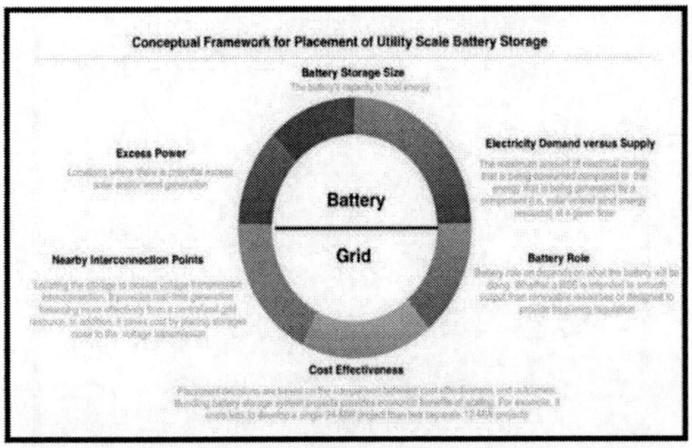

Figure 5. Conceptual Framework For Placement of Utility Scale Battery Storage

Our framework has two main dimensions; the first one is the factors' characteristic dimension, which has two main values: (1) battery and (2) grid. The second dimension is the factors that help in decision-making for the optimal location of battery storage. First, we searched in the literature, and the literature evaluation produced three factors that related to battery. Then we considered three more factors that related to grid. These factors assist in determining the optimal battery storages location for utilities. All six factors were developed from the background as is shown in Table II.

TABLE II. BATTERY & GRID RELATED FACTOR

Factor	Definition	Tech. Specification	Resource
Battery			
Battery Storage size	The battery's capacity to hold energy	Large centralized battery systems work better than smaller, distributed systems.	[23] [24]
Excess Power	Locations where there is potential excess solar and/or wind generation	Statistically significant areas using kernel density estimation (KDE) where there is high potential solar and/or wind generation	[25]
Electricity demand versus supply	The maximum amount of electrical energy that is being consumed compared to the energy that is being generated by a component (i.e., solar or/and wind energy resource) at a given time	The situation when energy supply is exceeding the demand	[26] [27]
Grid			
Nearby interconnecti on points	Locating the storage to closest voltage transmission interconnection . It provides real-time generation balancing more effectively from a centralized grid resource. In addition, it saves cost by placing storages close to the voltage transmission	Nearby 154-kV or 345-kV substations	[28]
Battery role	Battery role on depends on what the battery will be doing, such as whether a BSS is intended to smooth output from renewable resources or designed to provide frequency regulation.	Based on Table I "Energy Storage Technologies"	[24] [11]
Cost Effectivenes s	Placement decisions are based on the comparison between cost effectiveness and outcomes. Bundling battery storage system projects provides economic benefits of scaling. For example, It costs less to develop a single 24-MW project than two separate 12-MW projects.	Single centralized battery storage systems is preferred	[24]

B. *Artifact 2: GDSS Prototype*

This study proposes a GDSS solution to assist in the placement of battery storage systems by finding the optimal locations considering the electric grid constraints, the deployment requirements, and the potential benefits to the grid. This solution can aid in the battery storage systems location choices and provide actionable information for utilities, state-level decision-makers, and other stakeholders who are concerned about the grid reliability and the urgency of energy storage deployment as a distributed energy resource. Though a GDSS can provide a solution to address the placement of all types of energy storage systems, we chose to focus on Utility-scale storage to instantiate the conceptual framework and to demonstrate how an interactive, computer-based system can assist in decision-making considering the net and the potential benefits to the grid.

1) Data Selection and Acquisition

- SCE's DERiM Circuit Capacity Data: Southern California Edison's distributed energy resource interconnection map includes power electric lines and the capacity analysis in kilowatts by circuit line. The data is retrieved from DERiM web map.
- Solar Parcel Data: LA County is the data source for the solar parcel data. LA County solar map includes key data elements, such as: total roof area and area suitable for solar, potential solar system size, solar potential annual output, and potential cost savings.
- Electricity Household: Simply map is the source of data. The Electricity Household is in the average of the annual (yearly) consumption in dollar for each geographic region in 2016. Electricity is generally supplied by means of above or underground electric power lines.

2) Data Preparation Steps

- DERiM Circuit Capacity Data was loaded into ArcGis 10.3.1. The first layer was locating the SCE substations (837 stations) across LA County Figure 6.
- Considering the Nearby interconnection points requirement, substations with circuit capacity constraints with less than 150 kilovolts were excluded.

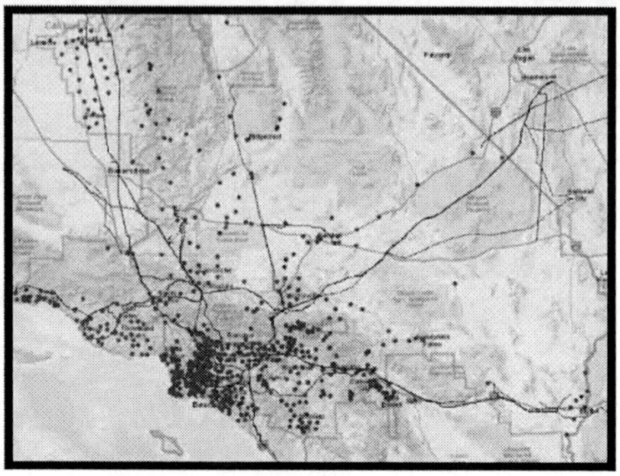

Figure 6. SCE Substations Across LA

- According to Sultan et al. [20], the solar rooftop's potential electricity output was calculated for each parcel by multiplying the rooftop's solar panel area, the solar panel yield, the annual average solar radiation on tilted panels (constant for LA county, equal to 2018.45), and the solar system's performance ratio which is the coefficient for losses (used default value = 0.75). Two fields were added in ArcMap attribute table to perform these calculations. Considering the V2G technology/exploiting excess power factor, these calculations are required to predict areas with potential solar excess generation assuming maximum adoption of solar rooftops in LA County [22].
- The electricity household data was added as an indicator for the electricity consumption level to serve the supply vs. demand factor Figure 7.

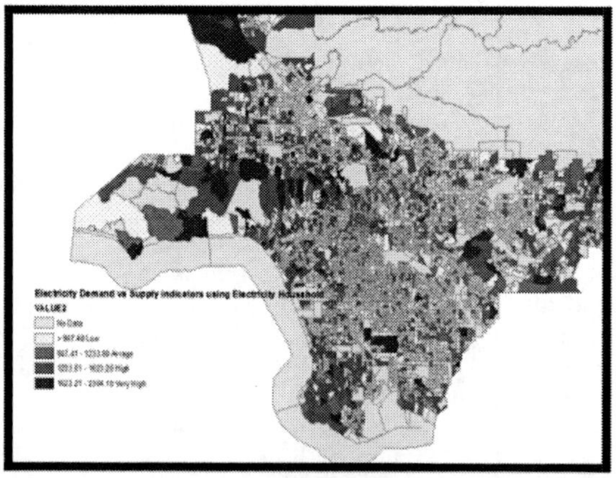

Figure 7. Electricity Consumption in LA County

- Potential Battery Location layer was created by joining the following map layers:
    - The excess parcel level power data based on step 3 to estimate the supply level.
    - The electricity household data to assess the

demand level. The assumption was made that less than 1000 in electricity household spending a year were considered low demand.

- The Substations data with the targeted requirements in step number 3 (more than 150 kilovolts). Then, buffer tool with 500 feet distance from the substations and locations that overlaps with low demand and high supply were identified.

3) *Analysis and Findings*

Our findings suggested 12 potential location for battery storage in SCE electrical substations based on the three spatial factors in the conceptual framework in Table II: 1) Demand vs. Supply 2) Nearby interconnection points 3) Excess power. The locations are listed in Table III and are shown in Figure 8.

Figure 8. The Suggested Battery Storage Locations

TABLE III. LIST OF LOCATION NAMES

ID	Substation Name	Substation Type
1	WALNUT	S -- Sub-transmission
2	ROSEMEAD	D -- Distribution
3	GOULD	S -- Sub-transmission
4	MESA	S -- Sub-transmission
5	LAGUNA	S -- Sub-transmission
6	BULLIS	D -- Distribution
7	CENTER	S -- Sub-transmission
8	CORNUTA	D -- Distribution
9	LIGHTHIPE	S -- Sub-transmission
10	SUNLIGHT P.T.	A -- District Pole Top
11	HASKELL	D -- Distribution
12	STADIUM	D -- Distribution

According to Jim Horstman, an industry consultant, pole top substation (Sunlight P.T.) included in Table III might not be used for batteries, as these are 'substation' built on a platform between poles. Thus, we have a final count of 11 potential locations for battery storage after excluding this one pole top substation.

Figures 9, 10, 11, and 12, show that substations of Stadium, Walnut, Rosemead and Gould are locations for placing battery storages. To illustrate, the yellow circles represent the location of the substation and the red circles represent the suggested location where there is excess power, supply in electricity is actually more than the demand level, and lastly that location is near a substation territory.

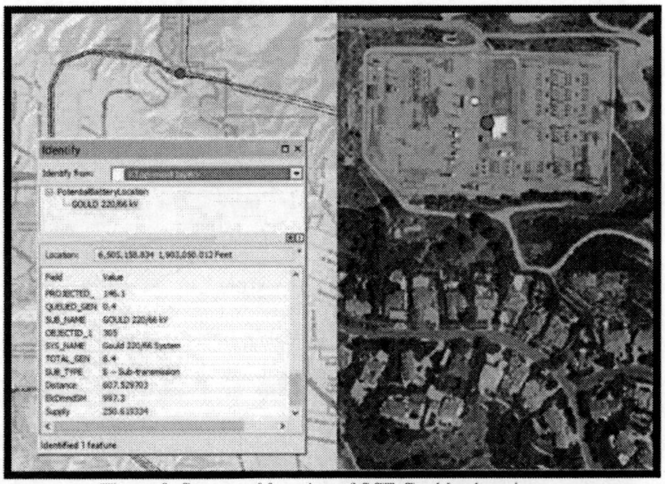

Figure 9. Suggested location of SCE Gould substation

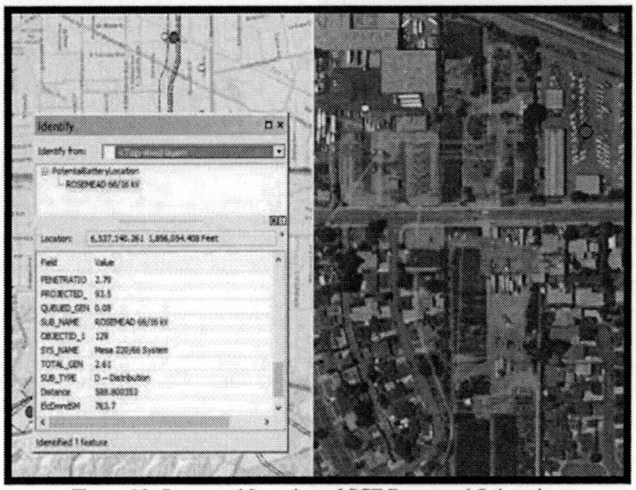

Figure 10. Suggested Location of SCE Rosemead Substation

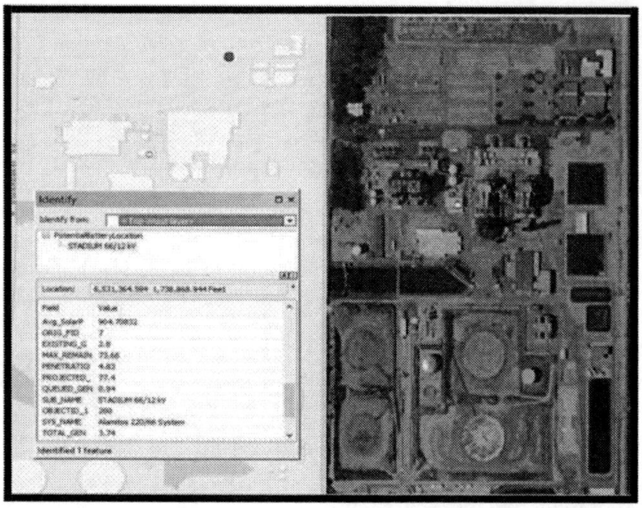

Figure 11. Suggested Location of SCE Stadium Substation

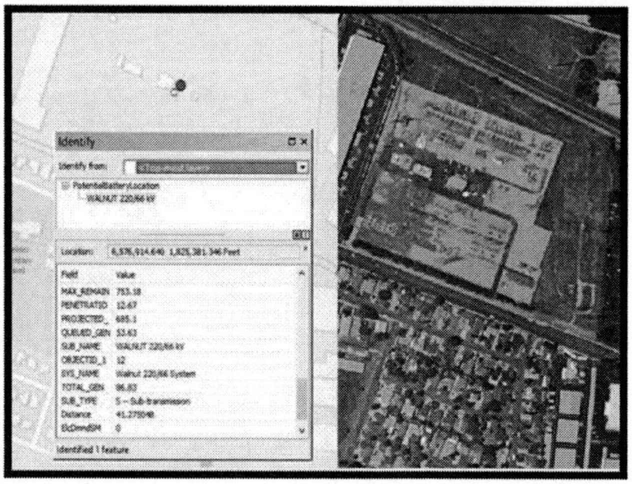

Figure 12. Suggested Location of SCE Walnut Substation

IV. EVALUATION

The project team applied a qualitative interview method to evaluate the prototype using socio-technical technique to assess the following metrics: propriety and utility. The researchers evaluated the different values held by the utilities by sending the following seven interview questions via email:

1. How useful is the conceptual framework (Fig.5.) in your view?
2. Is the framework complete from your perspective?
3. Do you agree with the factors' definitions in Table II?
4. Do you agree with the factors' technical specifications in Table II?
5. Do you see potential solution offered by the GDSS prototype (shown in Fig.8.)?
6. What changes would you recommend for improvement?
7. Do you see the potential in this research direction? In this case, both EV drivers' and utility executives' perspectives were part of the evaluation process to ensure that their unique stances were understood?

Feedback is currently being solicited from industry consultants and executives who can provide feedback from a utility perspective. Two participants responded with positive feedback and they highlighted the potential offered by the proposed artifact. According to Jim Horstman, a utility industry consultant, the conceptual framework is useful, complete, and it covers the relevant issues. Horstman mostly agree with the factors' definitions. However, he views the framework dimensions to be causing some confusion. For example, "the battery role is under grid rather than battery and the excess power and supply & demand are under battery when they would seem more appropriate to grid. Further Excess Power is a function of Supply and Demand (i.e., Excess Power = Supply > Demand) so not clear that is a separate issue" Horstman said. The industry consultant suggested another consideration relative to the nearby interconnection points, which is the space available for the battery storage. "While many substations might have ample space for the batteries others may have restraints" he said.

Horstman agreed with the technical specifications with the exception of including specific voltages (154 KV, 345 KV) as "those are utility specific e.g., SCE could be 220/500 KV for high voltage subs where it is stepped down to 133 KV, 66KV, etc.," The industry consultant recommended something more appropriate like high, medium voltage substations. In looking at the list in Table III, Jim Horstman saw that it includes sub-transmission and distribution substation that he believes would be categorized as medium and/or low voltage.

V. LIMITATION AND FUTURE WORK

The artifacts proposed in this paper aimed at providing a solution for decision-making to assist with the placement of battery storage systems by finding the optimal locations considering the electric grid constraints, the deployment requirements and the potential benefits to the grid. This could potentially save time and resources for developers and utility companies who are interested in the identification of optimal locations for the placement of storage systems.

Potential locations for battery storage can thus be prioritized in locations with excess power, where supply in electricity is actually more than the demand level, and where there is nearby a substation territory. However, a limitation of the GDSS prototype proposed in this paper is that it only deliberated three of the six factors of the suggested framework. Due to time and data availability constraints, the project team have chosen to build an artifact that addresses the placement of Utility-scale storage systems considering three of the six factors of the framework: 1) Demand vs. Supply 2) Nearby interconnection points 3) Excess power factor. We didn't address the battery size, the cost effectiveness and the battery role factors of the decision-making framework.

Another limitation is in the evaluation phase. It is important to recognize that the time constraint imposed limitations on the evaluation and what these limitations are.

The project team doesn't have the time to complete the evaluation phase and address the interviewees' comments after the solution delivery. The researchers don't have an opportunity to perform further iterations to improve the prototype. The project team needs to get and address the interviewees' suggestions to complete the research.

The utility and novelty of the solution is important to emphasize as the driving factors for this project. By developing a conceptual framework for decision-making that previously didn't exist, a great amount of time is reduced for utilities/grid operators who are interested in finding the optimal locations and the undertaking of integrating storage systems into the grid. The GDSS prototype can be regarded as an instantiation (to provide an instance of or concrete evidence in support) of the conceptual framework. So, it is important to realize that this prototype is only meant to serve as a good starting point for the illustration of how an interactive, computer-based system can assist in decision-making considering the net and the potential benefits to the grid.

In future DSR cycles, the project team will address all of the suggestions from interviewees after they get their feedback. The project team will evaluate the GDSS prototype many times through multiple iterations to improve its utility. The next iteration will offer a custom tool for developers/utilities, so that they are better able to evaluate the results. The prototype will run on a public server to give the research participants access to the application.

Moreover, the project team will add quantitative methods in the evaluation such as System Usability Scale (SUS) and cognitive walkthrough. The usability of this prototype will increase since both methods measure the usability of any application. The SUS and cognitive walkthrough will be administered to all research participants. In this case, the evaluation will involve more participants who will be given access to the application to ensure the validity of the evaluation results.

VI. CONCLUSION

This study aimed at addressing "How can we determine the optimal locations for energy storage for proper integration into the electricity system?" To answer the research question, we have developed a conceptual framework for decision-making that caters to utilities for it considers the impacts of the energy storage placement on the grid, and the electric circuit capacity constraints. The framework we offer in this paper is the first, to date, to address the research question. In addition, we have built a prototype of Geographic Decision Support Systems (GDSS), which is an elegant, interesting, and novel solution, to aid in the battery storage systems location choices and provide actionable information for utilities, state-level decision-makers, and other stakeholders who are concerned about the grid reliability and the urgency of energy storage deployment as a distributed energy resource.

From this research, we conclude that Battery Storage locations can be assessed geographically to minimize potential increases to overall electric system costs while still meeting customers' needs. Our solution provides evidence that GIS can play an integral role in the problem resolution.

REFERENCES

[1] Environmental Systems Research Institute. Enterprise GIS and the Smart Electric Grid. Available online: https://www.esri.com/library/whitepapers/pdfs/enterprise-gis-smart-electric-grid.pdf (accessed on 5 February 2017).

[2] Hirst, Eric and Brendon Kirby. 2000. Bulk-power basics: reliability and commerce. Available online: http://www.esper.com/hirst/ (accessed on 5 February 2017).

[3] Osborn, Julie, and Kawann, Cornelia. Energy Analysis Department Environmental Energy Technologies Division, Ernest Orlando Lawrence Berkeley National Laboratory, University of California Berkeley. *Reliability of the U.S. Electricity System: Recent Trends and Current Issues* (August 2001). Available online: https://emp.lbl.gov/sites/all/files/REPORT lbnl - 47043.pdf (accessed on February 18 2017).

[4] United States Department of Energy (2015). Quadrennial Technology Review: An Assessment of Energy Technologies And Research Opportunities: Chapter 3 - Enabling Modernization of the Electric Power System. Available online: http://energy.gov/downloads/chapter-3-enabling-modernization-electric-power-system (accessed on February 5 2017).

[5] The President's Council of Economic Advisers, the U.S. Department of Energy's Office of Electricity Delivery and Energy Reliability, and the White House Office of Science and Technology (August, 2013). Economic Benefits Of Increasing Electric Grid Resilience To Weather Outages. Available online: http://energy.gov/sites/prod/files/2013/08/f2/Grid%20Resiliency%20Report_FINAL.pdf (accessed on February 5 2017).

[6] California ISO (2014, December). Advancing and Maximizing the Value of Energy Storage Technology. Available online: https://www.caiso.com/Documents/AdvancingMaximizingValueofEnergyStorageTechnology_CaliforniaRoadmap.pdf (accessed on February 5 2017).

[7] Kamath, Haresh (2013, September). Environmental Systems Research Institute. Electricity Storage in Utility Applications. Available online: https://www.usea.org/sites/default/files/event-/EPRI Storage Presentation - USEA - 13 September 2013.pdf (accessed on February 22 2017).

[8] Takeda, H., Veerkamp, P., and Yoshikawa, H. Modeling design process. *AI magazine* 1990, *11*(4), pp. 37.

[9] Hevner, A. A three-cycle view of design science research. *Scandinavian Journal of Information Systems* 2007 ,19 (2), pp. 87-92.

[10] Tullis, Paul. Brilliant new strategies for storing power 2013. Available online: from http://www.slate.com/articles/health_and_science/alternative_energy/2013/03/energy_storage_technology_batteries_flywheels_compressed_air_rail_storage.html (accessed on February 5 2017).

[11] International Electrotechnical commission Market Strategy Board. White Paper: Grid Integration of large capacity Renewable Energy Sources and use of large capacity Electrical Energy Storage 2012. Available online: http://files.energystorageforum.com/IEC_Grid_Integration_of_Renewables_and_EES.pdf (accessed on February 5 2017).

[12] Greentech Media, and Energy Storage Association. Cleantech and Renewable Energy News and Analysis: U.S. Energy Storage Monitor Q4 2016. Available online: https://www.greentechmedia.com/research/subscription/u.s.-energy-storage-monitor (accessed on February 5 2017).

[13] United States Department of Energy 2013. Grid Energy Storage. Available online: https://energy.gov/sites/prod/files/2014/09/f18/Grid%20Energy%20Storage%20December%202013.pdf (accessed on February 5 2017).

[14] Bose, S., Gayme, D. F., Topcu, U., and Chandy, K. M. Optimal placement of energy storage in the grid. In *Decision and Control (CDC) 2012 IEEE 51st Annual Conference;* pp. 5605-5612. IEEE.

[15] Meneses de Quevedo, P., and Contreras, J. Optimal placement of energy storage and wind power under uncertainty. *Energies 2016,* 9(7), pp. 528.

[16] Fernández-Blanco, R., Dvorkin, Y., Xu, B., Wang, Y., and Kirschen, D. S. Energy Storage Siting and Sizing in the WECC Area and the CAISO System 2016.

[17] Xu, Y. *Optimal location of electrical energy storage in a power system with wind energy* 2014, Doctoral dissertation, Wichita State University.

[18] Resch, B. et al. GIS-based planning and modeling for renewable energy: Challenges and future research avenues. *ISPRS International Journal of Geo-Information* 2014, 3(2), pp. 662–692.

[19] Sultan, V., Bitar, H., and Hilton, B. Geographic decision support systems to optimize the placement of distributed energy resources. International Journal of Smart Grid and Clean Energy 2016, 5(3).

[20] Sultan, V., Alzahrani, A., Bitar, H., and Alharbi, N. Is California's Aging Infrastructure the Principal Contributor to the Recent Trend of Power Outage? *Journal of Communication and Computer* 2016, 13, pp. 225-233.

[21] Nazaripouya, H., Wang, Y., Chu, P., Pota, H. R., and Gadh, R. Optimal sizing and placement of battery energy storage in distribution system based on solar size for voltage regulation. In *Power & Energy Society General Meeting, 2015 IEEE* (pp. 1–5).

[22] Durlauf, S. N., and Blume, L. (Eds.). The new Palgrave dictionary of economics 2008. Volume 6, pp. 631-34. Basingstoke: Palgrave Macmillan.

[23] Jia, L., and Tong, L. (2016). Renewables and storage in distribution systems: Centralized vs. decentralized integration. IEEE Journal on Selected Areas in Communications, 34(3), pp. 665-674.

[24] Overton, Thomas W., and JD. Practical Considerations for Siting Utility-Scale Battery Projects 2016. Available online: http://www.powermag.com/practical-considerations-siting-utility-scale-battery-projects/ (accessed on February 2 2017).

[25] Chris Nelder, James Newcomb, and Garrett Fitzgerald 2016. Electric Vehicles as Distributed Energy Resources Rocky Mountain Institute. Available online: http://www.rmi.org/pdf_evs_as_DERs (accessed on October 2 2017).

[26] Sultan, V. "Exploring Geographic Information Systems To Mitigate America's Electric Grid Traffic Congestion Problem," Proceedings of the 4th International Symposium on Computational and Business Intelligence 2016, Switzerland, September 5-7.

[27] Sjödin, E., Gayme, D. F., and Topcu, U. Risk-mitigated optimal power flow for wind powered grids. In American Control Conference (ACC), 2012, pp. 4431-4437. IEEE.

[28] Overton, Thomas. Publication Of Electric Power Conference: Practical Considerations for Siting Utility-Scale Battery Projects. Available online: http://www.powermag.com/practical-considerations-siting-utility-scale-battery-projects/ (accessed on February 5 2017).

[29] The California Independent System Operator (ISO), the California Public Utilities Commission (CPUC), and the California Energy Commission (2014, December). Advancing and maximizing the value of Energy Storage Technology. A CALIFORNIA ROADMAP. Available online: https://www.caiso.com/Documents/Advancing-MaximizingValueofEnergyStorageTechnology_CaliforniaRoadmap.pdf (accessed on February 22 2017).

A Qualitative Risk Analysis of Smart Grid Security Protocols

Mustafa Saed
Electrical and Computer Engineering
University of Detroit Mercy
Detroit, USA
email: saedma@udmercy.edu

Kevin Daimi and Nizar Al Holou
College of Engineering and Science
University of Detroit Mercy
Detroit, USA
email: {daimikj, alholoun}@udmercy.edu

Abstract— **Recently, smart grids have been attracting enormous interest as the Internet continues to rapidly expand. A smart grid embeds Information Technology (IT) in its transmission networks to become two way communications. This was achieved by installing dedicated devices, such as a smart meters and related software. A smart grid provides efficient usage of electric power and energy-saving that is passed down to the consumer. Furthermore, in a smart meter, which is an essential component of a smart grid, an individual's behavior can be indirectly understood, for example, by examining the utilization status of electric power at homes. However, such handling of a smart meter is problematic in terms of privacy protection and other security concerns. To this end, this paper performs a risk analysis of a number of proposed security protocols for smart meters in smart grid networks. In this analysis, four different vulnerability types are identified. These vulnerabilities are augmented by four types of attackers and seven types of attacks that exploit those them. After assessing the likelihood and impact levels for all the different combinations of vulnerabilities, attacks and attackers, a risk matrix for various risk levels is exploited. Overall findings include nine low risks, six medium risks, and three high risks in some of the proposed security protocols for smart meters in smart grid networks. Finally, appropriate mitigation techniques for different risks are suggested.**

Keywords— Risk Analysis; Communication Protocols; Smart Meters; Smart Grid; Risk Matrix

I. INTRODUCTION

In order to operate efficiently, power generation and distribution companies need to predict future energy consumption. This prediction is typically based on historical household energy consumption patterns [1][2]. Accurate prediction is essential to reduce unnecessary power generation, leading to financial savings and reduced carbon dioxide emissions. Different time periods for collecting energy data result in different electricity costs. The tendency of consumers to reduce their power demand in response to high electricity prices [3] is a useful tool for demand side managers. In addition, the bidirectional communication capability of smart meters enables the monitoring and control of all household appliances to reduce energy consumption at the customer site [1]. The analysis and prediction of energy consumption involves processing large amount of data from networks of smart meters. A smart grid network is a special type of an ad hoc network where smart meters are needed. Such a network is very appealing in the modern era since it allows for many amazing applications, such as event correlation (network and substation level of the power distribution of smart grid) and scheduled load shedding [4]. For this new type of networks, classical secure communication protocols over the Internet infrastructures, such as Transport Layer Security (TLS) [5] and IPsec [6], may not be sufficient due to the dynamic topology of these networks. There are a number of contributions to protect the two-way direct and indirect communication of smart meters with collectors in smart grid through the introduction of two cryptographic protocols based on PKI. Nonetheless, introducing new protocols always increases the attacking surfaces.

In network security, risk analysis is a process of defining and analyzing the threats to the security protocols, as well as the infrastructures and the entities that these protocols operate on [8]. It is a combination of identifying potential vulnerabilities and attacker, and accessing impacts of the attacks. The goal of such an analysis is to produce a qualitative risk analysis document and optionally a mitigation plan that offers guidance to security architects and related business decision makers. The National Institute of Standards and Technology (NIST) provide guidelines on how to perform a risk analysis [9]. This will be followed in this paper.

Risk analysis and management is an important field of research in network security. There are mainly two ways to conduct such analysis, namely qualitative and quantitative.

The approach taken in this work is of the qualitative type. Some work on quantitative analysis is presented in [10][11]. As emphasized in [10], current industry standards for estimating cybersecurity risk are based on qualitative risk matrices. Lee et al. [12] performed an analysis of the risk of the bit-flipping attack, which might occur in LoRaWAN, a Media Access Control (MAC) protocol for Wide Area Networks, where one can change specific fields on ciphertext without decryption. Another interesting work is presented by Jacobson et al. [13] in which they dealt with risk analysis of a smart home automation system. Among all those efforts, the analysis in Cherdantseva et al. [14] is particularly interesting. The authors introduced risk assessment methods for SCADA systems related to the underlying DNPsec protocol to be used in secure communication protocols. The DNPSec is a security framework of Distributed Network Protocol Version 3 (DNP3) [15][16]. DNP3 is an open and optimized protocol developed for the Supervisory Control and Data Acquisition (SCADA) Systems supporting the utilities industries.

This paper attempts to facilitate the reduction of various risks associated with some proposed smart grid security protocols. To this end, a risk assessment method to comprehensively analyze a smart meter in smart grid and countermeasures for such risks is proposed. To achieve this, a list of vulnerabilities, attacks and attackers is created. This is followed by assigning the likelihood and impact of vulnerability, attack and attacker for each combination. Finally, a risk matrix is used to evaluate risks given the likelihood and impact levels. Once all the risks have been evaluated, a necessary next step is to address all the medium and high risks. For completeness, some recommendations for managing low risks for these protocols are also stated.

The reminder of the paper is organized as follows: Section II provides an overview of a number of recently proposed smart meter security protocols together with the security tools, techniques, and methods used by them. Section III reviews the types of vulnerabilities, attackers, and attacks attracted by these vulnerabilities. Section IV describes the risk analysis of the security protocols. The countermeasures of proposed security protocol risks are considered in detail in Section V. Finally, Section VI sets the conclusion and describes future work.

II. SMART GRID PROTOCOLS OVERVIEW

Smart grids subsystems and components can be protected and their security enhanced via cryptographic software and hardware, and other security techniques. In order to prepare the grounds for the risk analysis and mitigation methodologies introduced in the next sections, the security techniques, tools, methods and approaches followed by various researchers in the field of smart grid security will be briefly introduced. These will represent the input to the qualitative analysis addressed by this paper.

Saed et al. [7] presented security protocols for smart meters in smart grid. They proposed schemes for securing the direct and indirect smart meter-to-collector communications. The schemes are based on PKI. The authors proposed two different security protocols to enhance the security of the direct communication between smart meters and collectors in a smart grid. The first proposal secured direct communication without using the certificates and relied on public key cryptology. The second proposal protected the direct communication by using certificates and also depended on PKI. On both protocols, the substation is only directly connected to the collector. They further proposed an approach for the indirect communication between smart meters and collector. In this approach, the collector (gateway) should have initially received all the public keys and identities of the smart meters (user node). On the other hand, the smart meters should have the public key of the collector using any secure process. Furthermore, the predecessor and successor nodes are identified during installation and configuration of each smart meter. These protocols are design to provide secure communications among the three entities: a server (Substation), which is a supervisory node acting as a centralized authentication center or a Certificate Authority (CA); multiple center gateways (Collectors) that provide connectivity to the user nodes (smart meters); and multiple nodes that are essentially smart meters. The purpose of those communications is to allow a node (smart meter) to provide information, such as temperature readings, and electricity consumptions, to the gateways. To facilitate secure and authenticated communication between a node and a center gateway, the server acted as a Certificate Authority to authorize nodes and gateways. These protocols are expected to run over the DNPSec.

Dong et al. [17] proposed a protection scheme for the automation of smart grid system and patch distribution from the control center to data transmission security. Some of the functions were tested on the simulation platform through intrusion detection system and by using field devices, such as smart meter. Their proposal considered the security within smart meter but not for the smart meter communication, such as smart meter to smart meter and smart meter to collector [18]. Furthermore, their proposed protection system did not use digital signature to protect against forgery.

The sparse topology information of the smart grid was utilized by Giani et al. [19] to determine the attack meter sets. However, their work lacked the discussion of the system matrix acquisition. In fact, the design of the attack vector relied heavily on precise knowledge of the system matrix. In this case, it would not be easy to obtain such confidential information for an attacker who has limited access to the smart grid. Overall, a feasible unobservable attack scheme based on the incomplete system matrix has not yet been fully investigated. The authors in their proposal weren't covering the smart meter communication attack. They only mentioned for the possible vulnerabilities related to attack meter in physical layer.

Li et al. [20] presented an efficient and robust approach to authenticate data aggregation in smart grids. Aggregation refers to the communication between the smart meters and the collector. This is achieved via deploying signature aggregations, implementing batch verification, and signature amortization schemes to reduce communication overhead and number of signing and verification operations and provide fault tolerance. The authors proposed an efficient authentication scheme for power usage data aggregation in Neighborhood Area Networks (NAN) and smart meter to collector communications. The contributions for this work were represented by deploying digital signatures so that when the collector is out of service, alternative or backup collectors can execute the authentication approach without any additional configuration or setup. Their research also sought to reduce the number of signature and verification operations. However, the research is limited to authentication only. Thus, they are not securing the messages (readings) between smart meter and collector.

Many of the available schemes for both single-path and multipath routing are not suitable for meshed Advanced Metering Infrastructure (AMI) network [21]. Consequently, a security mechanism for multipath routing based on Elliptic Curve cryptology, digital signature, and Message Authentication Code (MAC) for such an AMI network was introduced. This approach allowed the Certificate Authority to execute a lot more work than it normally should do (issuing certificates). The extra load included controlling the nodes' creation of public and private key. In this scheme, nodes (smart meters) performed a number of computations despite their known limited computing power. This also tended to slow the system. Furthermore, having a smart meter sending its information to all the neighboring smart meters with no protection at all would introduce an immense threat. This provides a potential attacker the opportunity for attacking more than one target (smart meter) as they all have the information of the source meter. Therefore, the neighboring nodes acted as intermediate nodes, and consequently performed more calculations and broadcasting of the results. This means all other nodes (smart meters) have now the information. This implies, there are many nodes that the attacker can try and many nodes will be affected.

Yan et al. [22] introduced an interesting security protocol for AMI communications in smart grid where the smart meters are interconnected through wireless network. Their techniques indicated that the Public Key Infrastructure (PKI) is not desirable and relied on symmetric key cryptology instead. However, the number of symmetric keys used is large and possibly comparable to the number of keys should PKI have been followed. Furthermore, smart meters have limited capabilities, and therefore, verifying the MAC should have been left to the collector. The authors did not specify what would happen when the two MACs are not equal. This implies that the integrity of a meter's reading is not handled correctly.

Seo et al. [23] discussed the use of public key infrastructure (PKI) in smart grid and what security requirements need to be implemented in smart grid architecture including the smart meter to secure the smart meter communication in the AMI. The authors did not propose any security technique/protocols to secure the smart grid network but only provided a survey.

Zhao et al. [24] provided the fundamental limit of cyber-physical security in the presence of low sparsity unobservable attacks. It is shown in [25][26] that a complete system matrix can be identified using an independent component analysis method. Nevertheless, such attack schemes might not be easy to implement as all meter data are required to be known and all the meters are required to be controlled. On the other hand, several detection and defense schemes are provided based on the complete knowledge of the system matrix. The off-line method, based on the Kullback-Leibler distance, is proposed to track malicious attacks using historical data [27].

A distributed incremental data aggregation approach, in which data aggregation is performed at all smart meters involved in routing data from the smart meter to the collector unit, was introduced by Li et al. [28]. In this research, the authors presented an efficient information aggregation approach, in which an aggregation tree, constructed via breadth-first traversal of the graph and rooted at the collector unit, is deployed to cover all smart meters in the neighborhood. This protocol can let the control unit collect all smart meters' information in the area. Furthermore, to protect users' privacy, all information is encrypted by a homomorphic encryption algorithm. Since no authentication scheme is emphasized, the approach faces the potential risk that malicious smart meter can forge packets, thus causing the smart grid system to fail to detect or diagnose bogus data. Adversaries can maliciously forge their own data to manipulate the aggregation results.

TABLE I. LIST OF ACRONYMS AND SYMBOLS

Acronyms/Symbols	Meaning
A1	MITM
A2	Impersonation
A3	Single Point of Failure
A4	Key Escrow
A5	Cryptanalysis and Quantum Computer
A6	Forward Secrecy
A7	Downgrade Attack
CA	Certified Authority
DNP3	Distributed Network Protocol Ver.3
DNP3Sec	DNP3 Security
IKE	Internet Key Exchange
LoRaWAN	MAC Protocol for WAN
MAC	Media Access Control
MITM	Man-in-the-Middle
NIST	National Institute for Standard & Technology
OEM	Original Equipment Manufacture
PKI	Public-Key Infrastructure
QoS	Quality of Service
SCADA	Supervisory control and Data Acquisition
T1	Clover Outsider
T2	Knowledgeable Insider
T3	Non-Profit Organization
T4	For-Profit Organization
TLS	Transport Layer Security
V1	Lack of Authenticity
V2	Centralized Topology
V3	Weak Cryptography
V4	Misuse of Public Key Cryptography
WAN	Wide Area Network

III. TYPES OF VULNERABILITIES, ATTACKS AND ATTACKERS

In this section, the relevant building blocks of risk analysis are presented. First, the risk matrix, which determines the risk level given the likelihood and impact levels of an attack being carried out by an attacker are introduced. Then, the different types of attackers, vulnerabilities and attacks are depicted. Some of the vulnerabilities presented here may not be applicable to these protocols but are otherwise common for other networks than smart grid. Table I presents a list of acronyms and symbols used in this paper.

A. Security Concerns

There are several security concerns for these protocols. The list below describes the completeness:

1) Loss of sensitive data: This could mean either user's personal data, or server/gateways (Substation/Collector) private data, or even statistical data that should be kept secret. Those data could be ephemeral, such as one-time session keys, or could have long term impact, such as user's credit card information or social security number.

2) Financial loss: This could indicate that the user is overcharged for services that she/he did not receive. For gateways (collectors), this could mean that the gateway did not receive the credit for the service it provided. It could also mean a malicious modification to financial data at the gateway, which incurs financial loss.

3) Denial of service (DoS): Certain attacks are able to disable partial or full part of the smart grid network, so that nodes (smart meters) do not receive services from gateways (collector). This kind of attacks may be localized, such as unauthorized access to specific smart meter at the same domain, or may also be global, such as unauthorized access to specific smart meter from different domain. Depending on the type of attack, DoS may last for a short or a long period. In addition, damage to the hardware is also a security concern here. However, this paper focuses on evaluating the secure communication protocol. Therefore, such a concern is beyond the scope of this paper.

B. Types of Vulnerabilities

Below different types of vulnerabilities are presented:

1) Lack of authenticity (V1): authenticity is missing in almost half of the protocol, namely, Section A of [7]. For the other half of the scheme, a certificate is used.

2) Centralized topology (V2): the protocol uses a centralized structure, where a single server (substation) is responsible for handling enrollment and certificates for all collectors and smart meters.

3) Weak cryptography (V3): A protocol may employ a weak cryptography that is vulnerable to cryptanalysis, or it may employ a strong cryptography that is secure against cryptanalysis today, but will be broken in the future. An example of the first case is SHA-1 hash function [29], and an example of the second case is RSA [30] or ECC [31] against cryptanalysis using quantum computer in the future. For the protocol to be analyzed, the underlying cryptography primitives are not specified [32]. That it deploys RSA or ECC based solutions is assumed, as well as SHA2 [33] or SHA3 [34] functions at a desired security level. This assures the protocol to be robust against today's cryptanalysis, but still render to the vulnerability of quantum cryptanalysis in future.

4) Misuse of public key cryptography (V4): the protocol uses public key cryptography to establish secure communication channels between entities. However, in modern cryptography [35], public key cryptography is usually used to establish a session key, rather than used directly for communication, in order to provide additional security features, as well as improved performances.

C. Types of Attacks

Attacks that exploit above mentioned vulnerabilities are described here:

1) Man-in-the-Middle (MITM) (A1)

MITM attacks are one of the most classical attacks in cryptography and network security. In Moore [36], the author gives a tutorial of MITM attacks. In terms of the MITM attacks against analyzed protocol, the following are observed:

a) Attacking strategy: the attacker secretly relays and possibly alters the communication between two parties, such as smart meter to collector who believe they are directly communicating with each other.

b) Assumptions: it is reasonable to assume that the attacker is able to passively eavesdrop the communication between two entities; it is however hard or infeasible for the attacker to break the authentication within real time.

c) Common vulnerability for this attack: lack of authentication methods, for examples, certificates and/or pre-shared keys.

d) Consequences: the attacker makes independent connections with the victims and relays messages between them to make them believe they are talking directly to each other over a private connection, when in fact the entire conversation is controlled by the attacker. The attacker must be able to intercept all relevant messages passing between the two victims and inject new ones.

2) Impersonation (A2)

Impersonation is another classical attack in cryptography and network security [37]. The secure communication protocols with impersonation attack are analyzed as follows:

a) Attacking strategy: the attacker claims to be someone else, a legitimate user (smart meter) or a substation node.

b) Assumptions: the attacker is able to get the public keys and IDs from all entities (smart meter, collector and substation); but has no access to the secret keys.

c) Common vulnerability for this attack: lack of authentication methods, for examples, certificates and/or pre-shared key.

d) Consequences: the attacker convinced entities that he/she is a legitimate owner of an ID.

3) Single Point of Failure (A3)

Single point of failure attack is common in network securities [38]. The proposed protocols in [7] are analyzed with this attack due to the centralized network topology that the protocol employs:

a) Attacking strategy: the server (substation) is a single point of failure. The attacker focuses its resources to attack this single point rather than the whole smart grid system.

b) Assumptions: the attacker is able to break into the server (substation).

c) Common vulnerability for this attack: having a centralized structure

d) Consequences: Total compromising of the protocol.

4) Key Escrow (A4)

Key escrow attacks, as described in OH et al. [39], is a common attack in cryptography as observed in the following:

a) Attacking strategy: the server (substation) is responsible for authentication, so it is able to authenticate a fake user (smart meter) or revoke a legitimate user.

b) Assumptions: the server (substation) is malicious, or is compromised by the attacker

c) Common vulnerability for this attack: have a centralized structure

d) Consequences: Total control of the protocol

5) Cryptanalysis and Quantum Computing (A5)

This attack exploits the weakness in cryptography, using cryptanalytic methods, such as Shor's algorithm [32]:

a) Attacking strategy: use cryptanalytic tools (with quantum computers [20], if necessary) to break the existing cryptosystem.

b) Assumptions: the underlying cryptosystem is vulnerable to cryptanalysis and quantum computers.

c) Common vulnerability for this attack: RSA [30] and ECC [31] are both vulnerable to quantum computers.

d) Consequences: Total capturing of the protocol.

6) Forward Secrecy (A6)

Forward secrecy is a notion associated with network security and secures protocol designs [40]. The proposed protocols in [7], have the following properties:

a) Attacking strategy: once the attacker gains control of a session (through other attacks, for example, MITM), the session key is used to learn previous and future keys.

b) Assumptions: the attacker is able to learn the secret information of at least a single session.

c) Common vulnerability for this attack: a bad key update schedule; usage of statistical keys; lack of short term (one time) keys.

d) Consequences: the attacker steals secret information of entities (smart meter, collector and substation) causing all previous/future data to be at risk.

For completeness, the following attack is also presented. Currently, this is not applicable to the protocols in [7]. When these protocols will evolve in the future, and there will be more than one version available for use, this attack becomes applicable.

7) Downgrade Attack (A7)

This type of attack exploits the fact that some earlier version of the protocol uses weaker cryptography. For historical and legacy reasons, the current protocol needs to be able to communicate with those earlier versions [41].

D. Types of Attackers

It is important to model the attackers, as different attackers have different likelihood to launch attacks, and the consequences, even for a same attack, may vary for different attackers. Therefore, four types of attackers are considered in this risk analysis [42].

1) Clever Outsider (T1): Examples include a high school student, hacker, and researcher. Those types of attackers are usually limited to their knowledge of the underlying cryptography and the topology of the network. They are also likely to be constrained by the hardware they can access. For example, they are not likely to be able to launch attacks from multiple computers in parallel. In most cases, they are honest but curious. It implies that they will only exploit vulnerabilities that are exposed to them; they will not actively look for vulnerabilities. The goal of their attack is usually financial gains or publicity.

2) Knowledgeable Insider (T2): Examples include a disgruntled employee. Unlike outside attackers, inside attackers are much more knowledgeable of the required skills to launch attacks. They are also more capable of identifying critical point of the network in order to maximize the impact of the attack. However, inside attackers work alone, as they are not organized (c.f T3 and T4) and do not want to be identified. Therefore, just like T1 attackers, they are also likely to be constrained by the hardware they can access. In addition, it is safe to assume that they are malicious. They know about all the vulnerabilities of the protocol. The goal of their attack is usually financial gains and vengeance.

3) Non-Profit Organization (T3): Examples include research groups, and collaborators on the Internet. Those are potentially at large scale organized groups,. Therefore, these groups consist of experts in the related area. Since they are large scale organizations, they are able to launch distributed and parallel attacks. As non-profit organizations, the goal of their attack is usually research or charity oriented publicity.

4) For-Profit Organization (T4): Examples include a competitor Original Equipment Manufacturer (OEM) and a tier-1 supplier. Similar to T3 attackers, being large organizations means they have access to all sorts of resources related to the attack, including both the required skills and knowledge, and necessary equipment's. In addition, since they are profitable organizations, it is also possible for them to hire/buy additional resources to maximize the impact of their attacks. Those attacks are usually profit-oriented.

IV. RISK ANALYSIS

In this section the risk analysis of the security protocols will be described.

A. *Risk analysis metrics*

In carrying out the risk analysis, it is important to de-couple the assessments of likelihood from that of impact, otherwise the same factor would be counted twice. However, doing this, in general, is rather difficult. The likelihood of an (vulnerability, attack, attacker) combination is assessed, and only look at factors like, for the level of difficulty needed by the attacker to exploit the vulnerability, and if the attacker requires some special tools/knowledge. On the other hand, when assessing the vulnerability of the combination (vulnerability, attack, attacker), it is already assumed that the exploitation of the vulnerability is possible, and then try to determine the impact in terms of loss in Quality of Service (QoS) or financial impact. The following metric are adopted in this paper:

1) Low: Assigned when compromising a small part the network, and incurring minimal or no financial loss.

2) Medium: Allocated when compromising a large part of or the whole network for a limited time, and incurring some financial loss;

3) High: Vilified when compromising a large part or the whole network for a very long time or compromise of sensitive information like private keys, credit card numbers, and incurring significant financial loss.

Table II shows the risk matrix that maps a (likelihood, impact) combination to a risk, all of them have three levels: low, medium and high. Throughout this paper, this table will be used to determine the risk level [43].

TABLE II. RISK MATRIX

Impact	Likelihood		
	Low	*Medium*	*High*
Low	Level=Low	Level=Low	Level=Medium
Medium	Level=Low	Level=Medium	Level=High
High	Level=Medium	Level=High	Level=High

B. *Detailed risk analysis*

In the following subsections, a detailed risk analysis of every possible combination of vulnerabilities, attacks and attackers is carried out. A summary of the risk analysis is presented in Table III. It is worth mentioning that there are 112 combinations (4 vulnerabilities x 7 attacks x 4 attackers). As different combinations of vulnerabilities, attacks and attackers, vulnerabilities are not independent of attacks example, the combinationV1 and A3 is not a valid one since a single point of failure attack cannot exploit the protocols lacking authenticity vulnerability. This paper will concentrate on the most common combinations.

1) (V1, A1, T1): To carry out an MITM attack, an attacker requires the knowledge of the underlying cryptography, network topology, and protocol design. A clever outsider is unlikely to possess all this knowledge. For example, if the attacker does not know the topology of the network, they cannot easily identify a gateway (collector), or the link from a smart meter to the collector. Therefore the likelihood is assigned to be low. In terms of impact, a clever outsider is unlikely to compromise more than one segment of the smart grid network at a time, which would incur minimal financial loss. Here, the impact to be also low. Hence, the risk is low.

2) (V1, A1, T2): As stated in IV-B-1, this attack requires the knowledge of the underlying cryptography, network topology, and protocol design. A knowledgeable insider is likely to possess some, if not all, of this knowledge. The likelihood is assessed to be medium. Just like IV-B-1, a knowledgeable insider is unlikely to compromise more than one segment of the smart grid network at a time, which would incur minimal financial loss. So, the impact will be stated low. Accordingly, the risk is low.

3) (V1, A1, T3): Compared to IV-B-2, a non-profit organization is also likely to possess some, if not all, knowledge of the underlying cryptography, network topology and protocol design. However, unlike IV-B-4, the motivation for such an attacker is not strong. Such an attacker is mainly interested in personal gains, such as producing research papers or personal publicity. Therefore, the likelihood is stated to be medium. In terms of impact, a non-profit organization has the capability to compromise a large part of the smart grid network at a time, which would incur a non-negligible financial loss. The impact is assessed to be also medium. Hence, the risk is medium.

4) (V1, A1, T4): A for-profit organization is likely to possess all the required knowledge for this attack. In addition, the motivation of such an attacker is quite strong. Usually, such an attacker will have financial interest and brand reputation. Therefore, the likelihood is high. Just like IV-B-3. A for-profit organization has the capability to compromise a large part of the smart grid network at a time, which would incur a non-negligible financial loss. Consequently, the impact is set to medium. Hence, the risk is high.

5) (V1, A2, T1): Similar to an MITM attack, an impersonation attack, require an attacker to be knowledgeable of the underlying cryptography and protocol design. A clever outsider is unlikely to possess all this knowledge. This implies the likelihood should be low. The impact of this attack is also similar to that of MITM attacks. A clever outsider is unlikely to compromise more than one segment of the smart grid network at a time. A minimal financial loss is expected. The assessment of the impact is low. Hence, the risk is low.

6) (V1, A2, T2): A knowledgeable insider is likely to possess some, if not all, of required knowledge, so, the likelihood is medium. Just like IV-B-5, a knowledgeable insider is unlikely to compromise more than one segment of the smart grid network at a time, which would incur minimal financial loss, Therefore, the impact is assessed to be low. Hence, the risk is low.

7) (V1, A2, T3): A non-profit organization is likely to possess some, if not all, of the knowledge of underlying cryptography, as well as the protocol design. Nonetheless, the motivation of such an attacker is not strong, since the attacker is mainly interested in publicity gains rather than financial gains. So, the likelihood is assessed to be medium. In terms of impact, a non-profit organization has the capability to compromise a large part of the smart grid network at a time, which would incur a non-negligible financial loss.The impact is stated to be also medium. Hence, the risk is medium.

8) (V1, A2, T4): As stated in IV-B-5, this attack requires the knowledge of the underlying cryptography, protocol design. A for-profit organization possesses this knowledge. On the other hand, for the motivation, is strong for such an attacker, due to potential financial gains or brand reputation gains. Therefore, the likelihood of such an attack will be high. In the meantime, a for-profit organization has the capability to compromise a large part of the smart grid network at a time, which would incur a non-negligible financial loss. The impact is assessed to be medium. Hence, the risk is high.

9) (V1, A3, T1): This protocol has a single point of failure, a centralized server (Substation), that is responsible for authentication. An attacker can try to take it offline to cause disruption in service. Alternatively, if he/she can break the server (Substation), he/she can authorize unauthorized nodes, (smart meter, collector) or revoke authorized nodes (smart meter, collector). However, such an attack would require access to points of failure, which a clever outsider is unlikely to possess. The likelihood is assessed to be low. In terms of impact, if the attacker is able to launch this attack, then there will be significant disruption in smart grid service. However, it is unlikely that this attack alone would lead to loss of sensitive information. So the impact is assessed to be medium. Hence, the risk is low.

10) (V2, A3, T2): Unlike IV-B-9, a knowledgeable insider will likely have access to the topology of the network. This will help him/her to gain access to the server (substation), which is a single point of failure. In addition, this attack does not require a lot of internal knowledge or resources. The likelihood is assessed to be high. For the impact of the attack, there is not much difference between this attack and that of IV-B-9. Hence, the impact is assessed to be medium. Hence, the risk is high.

11) (V2, A3, T3): Just like IV-B-9, a non-profit organization is unlikely to have access to points of failure, if there are any. Such an access requires inside knowledge of the smart grid network. Therefore, the likelihood is assessed to be low. Compared to IV-B-10, it is quite interesting to notice that a more capable attacker has a lower likelihood to launch this attack. Similar to IV-B-9 and IV-B-10, a non-profit organization will be able to significantly disrupt the service of the smart grid via this attack, without causing losses of sensitive information. As a result, the impact is assessed to be medium. Hence, the risk is low.

12) (V2, A3, T4): Even though a for-profit organization may not have easy access to points of failure, such attackers usually have more resources available to them than a nonprofit organization. In addition, the motivation of a for-profit organization is stronger than all other three types of attackers. The likelihood is assessed to be medium. The impact of this attack remains medium for the same reason stated before. Hence, the risk is medium.

13) (V2, A4, T1-T4): The proposed security protocols in [7] have assumed that the server (Substation) is always trusted, so the likelihood is assessed to be not applicable (N/A). In terms of impact, if such an attack is successful then there will be a non-negligible financial loss. The impact is assessed to be medium. Still, the risk is (N/A).

14) (V3, A5, T1-T2): Modern cryptographic schemes have strong mathematical foundations and are usually designed to be secure for the foreseeable future. Cryptanalyzing these schemes is extremely difficult, if not impossible. The likelihood is assessed to be low for now. With the recent research on quantum computers, it is possible that there will be usable quantum computers in the next couple of decades [44], which would mean that most of the existing cryptographic schemes could be broken. However, such a quantum computer, if it did exist, wouldn't be easily accessible to a clever outsider or a knowledgeable insider. The likelihood is assessed to be medium for future. In terms of impact, if this attack is successful, the attacker will have access to sensitive information and there will most likely be significant financial loss, so the impact is assessed to be high.

15) (V3, A5, T3-T4): Just like IV-B-14, cryptanalyzing schemes is extremely difficult, if not impossible, even for large organizations. The likelihood is assessed to be low for now. If and when there will be a quantum computer, a large organization would be able to get access to it. The likelihood is assessed to be high for future. Just like IV-B-14, if this attack is successful, the attacker will have access to sensitive information to the smart grid network and there will most likely be significant financial loss, so the impact is assessed to be high.

16) (V4, A6, T1): To carry out this attack, an attacker requires the knowledge of the underlying cryptography, as well as the protocol design. A clever outsider is unlikely to possess all this knowledge. In addition, this attack is meaningful only if some other attack like MITM is also successful, so the likelihood is assessed to be low. In terms of impact, a clever outsider is unlikely to compromise more than one segment of the smart grid network at a time, which would incur minimal financial loss. The impact is assessed to be also low. Hence, the risk is low.

17) (V4, A6, T2): A knowledgeable insider is likely to possess some, if not all, knowledge of the underlying cryptography and the protocol design. However, as stated in IV-B-16 this attack requires some other attack, such as the MITM to be also successful. Therefore, the likelihood is assessed to be low. In the meantime, a knowledgeable insider is unlikely to compromise more than one segment of the smart grid network at a time, which would incur minimal financial loss. The impact is assessed to be low. Hence, the risk is low.

18) (V4, A6,T3): A non-profit organization is likely to be knowledgeable about cryptography and network design. However, the likelihood is assessed to be low for two reasons. First, as stated earlier, this attack is dependent on other attacks. Secondly, the motivation of such an attacker is not strong, since such an attacker is mainly interested in research publications or personal publicity. In terms of impact, a non-profit organization has the capability to compromise a large part of the smart grid network at a time, which would incur a non-negligible financial loss. The impact is assessed to be also medium. Hence, the risk is low.

19) (V4, A6, T4): The likelihood for a for-profit organization to launch this attack is medium. Such an attacker maintains the required knowledge. It is also motivated since a successful attack will lead to financial gains. However, this attack depends on other attacks which reduce the likelihood of this attack. Just like IV-B-18, a for-profit organization has the capability to compromise a large part of the smart grid network at a time, which would incur a non-negligible financial loss. The impact is assessed to be medium. Hence, the risk is medium.

TABLE III. RISK ANALYSIS SUMMERY

Attack	Risk Analysis		
	Likelihood	*Impact*	*Risk*
(V1,A1,T1)	Low	Low	Low
(V1,A1,T2)	Medium	Low	Low
(V1,A1,T3)	Medium	Medium	Medium
(V1,A1,T4)	High	Medium	High
(V1,A2,T1)	Low	Low	Low
(V1,A2,T2)	Medium	Low	Low
(V1,A2,T3)	Medium	Medium	Medium
(V1,A2,T4)	High	Medium	High
(V2,A3,T1)	Low	Medium	Low
(V2,A3,T2)	High	Medium	High
(V2,A3,T3)	Low	Medium	Low
(V2,A3,T4)	Medium	Medium	Medium
(V2,A4,T1-T4)	N/A	Medium	N/A
(V3,A5,T1-T2)	Low	High	Medium
(V3,A5,T3-T4)	Low	High	Medium
(V4,A6,T1)	Low	Low	Low
(V4,A6,T2)	Low	Low	Low
(V4,A6,T3)	Low	Medium	Low
(V4,A6,T4)	Medium	Medium	Medium

V. RISK MITIGATIONS

The countermeasures of proposed security protocol risks are considered in detail in this section.

A. Mitigations of Medium and High Risks

1) (V1, A1, T3-T4), (V1, A2, T3-T4): A straightforward way to mitigate MITM and impersonation attacks is to use digital certificates, such as Public-Key Infrastructure (PKI) and secure the channel using protocols like Transport Layer Security (TLS) [5]. With authentications, when a sender sends packages to a receiver, the receiver will check the authenticator associated with the sender and the package. This can be done via a digital signature for the sender's identical, signed by a Certificate Authority, and a MAC that binds the identity, the signature and the package. In this case, if one wishes to launch a MITM attack or an impersonate attack, he or she will have to break the underlying cryptography, which is prohibited by the given vulnerability, attack and attacker combinations.

2) (V2, A3, T2), (V2, A3, T4): Forthright way to mitigate vulnerabilities like the single point of failure is to decentralize the server by replication of resources and sharing the cryptographic key materials. This effectively stops attacks on single point of failures as even if one server is compromised, there are still adequate number of servers remains to provide required functionalities.

3) (V3, A5, T1-T4): To mitigate the medium current risk and possibly high risk in future, the use of cryptographic schemes that have at least 128-bit security is recommended (see, for example, NSA's suite B Cryptography Standards [45]). Standardized 128-bit secure cryptography is believed to be robust against any existing cryptanalysis. The number of operations to break the cryptography is over 2128 bit operations, which is beyond the capability of classical computers. A even better solution, nonetheless, is to use cryptographic schemes that are believed to be quantum-safe, such as New Hope Key exchange algorithms [46], NTRU cryptosystems [47], in a hybrid mode [48]. The hybrid mode means one uses a quantum-safe scheme, for example, NTRU, in parallel with a classical scheme, such as Diffie-Hellman key exchanges. With a right configuration, the system will be as strong as the stronger scheme of the two. This provides sufficient security against classical attackers today, as well as potential quantum attackers in future.

4) (V4, A6, T4): To mitigate the risk of forward secrecy attacks, the use of ephemeral keys instead of static keys are recommended. Connections via ephemeral keys are sort of a de facto method to design secure communication protocols, such as TLS [5] and IKE [49]. In those protocols, the session key that was used for communications are derived from a long time authenticated key and an ephemeral key generated at run time. Therefore, for each session, the key is different and independent from all previous ones. As a result, a leakage of either a session key or the long time authenticated key will only have localized effect. The attacker will learn information about the particular session only, not the entire sessions that the user has participated.

B. Mitigation of Low Risks

In principle, there isn't really any need to address the low risks in the system right away, but it is important to keep an eye on them, as those risks may become medium or high in future depending on several parameters like scientific, technological, and algorithmic developments. An example is that of cryptanalysis attacks using quantum computers, whose likelihood is rated as low at the moment because quantum computers don't exist, but there is a non-negligible chance that quantum computers will become viable in the next decade or so, and therefore, the likelihood is rated as medium for future. The mitigations of low risks are given for the sake of completeness. It is worth noting that their mitigations are usually very similar to the medium/high risks in the same attack/vulnerability category. In a bit more details, (V1, A1, T1), (V1, A1, T2), (V1, A2, T1), and (V1, A2, T2) can be mitigated using techniques similar to Section IV-A-1. Those attacks rely on the lack of authenticity. As discussed earlier, with authenticators, one can effectively check the integrity and the authenticity of the packages it receives, and therefore, defeats those attacks. (V2, A3, T1) and (V2, A3, T3) can be mitigated using techniques similar to Section IV-A-2. They rely on single point of failure. So using a decentralized topology safely is assumed these attacks will fail. Similarly, using ephemeral keys will effectively stop attacks in (V4, A6, T1-T3) as shown in Section IV-A-4. With ephemeral keys, the leakage of a single key does no longer imply the leakage of all keys as suggested in.

VI. CONCLUSION

Security protocols for the smart grid are designed and implemented to protect the communications between various components within the grid. If these protocols reveal weaknesses, devastating consequences could take place. Therefore, these protocols should be fully analyzed and studied to ensure they provide robust security. In an attempt to participate in the effort of ensuring strong security implementation, this paper provides a framework for analyzing qualitative risk in smart grid security protocols. Risk factors are extracted together with their related issues. This analysis concluded nine low risks, six medium risks, and three high risks in these protocols. Countermeasures are proposed, and appropriate mitigation techniques for the identified risks are suggested.

REFERENCES

[1] S. Depuru, L. Wang, and V. Devabhaktuni, "Smart meters for power grid: Challenges, issues, advantages and status," Renewable and sustainable energy reviews, vol. 15, no. 6, pp. 2736–2742, 2011.

[2] S. Finster and I. Baumgart, "Privacy-Aware Smart Metering: A Survey," IEEE Communications Surveys & Tutorials, pp. 1088–1101, 2014.

[3] K. Herter, P. McAuliffec, and A. Rosenfeld, "An exploratory analysis of california residential customer response to critical peak pricing of electricity, " Energy, pp. 25–34, 2007.

[4] Future-Proofing Smart Grid Infrastructure Meter Designs. [Online].Available: https://www.microsemi.com/applications/industrial-m2m-wireless/smart-metering, [retrieved: March 2018].

[5] T. Dierks and E. Rescorla, "The Transport Layer Security (TLS) Protocol," Version 1.2. Internet Engineering Task Force, 2008. [Online]. Available: https://tools.ietf.org/html/rfc5246, [retrieved: March 2018].

[6] S. Kent and R. Atkinson, "IP Encapsulating Security Payload (ESP)," Internet Engineering Task Force, 1998. [Online]. Available: https://tools.ietf.org/html/rfc2406, [retrieved: March 2018].

[7] M. Saed, K. Daimi, and N. Al-Holou, "Approaches for Securing Smart Meters in Smart Grid Networks," International Journal on Advances in Systems and Measurements, vol. 10 no. 3&4, pp. 265-274, Dec. 2017.

[8] W. Stallings, "Network Security Essentials - Applications and Standards," (4. ed., internat. ed.). Pearson Education, 2010.

[9] NIST, "Guide for Conducting Risk Assessments," National Institute of Standards and Technology, 2012. [Online]. Available: http://doi.org/10.6028/NIST.SP.800-30r1, [retrieved: March 2018].

[10] L. Allodi, W. Shim, and F. Massacci, "Quantitative Assessment of Risk Reduction with Cybercrime Black Market Monitoring," In 2013 IEEE Symposium on Security and Privacy Workshops, pp. 165–172, May 2013, CA, USA.

[11] L. Allodi and F. Massacci, "Security events and vulnerability data for cybersecurity risk estimation," International Journal on Risk Analysis, vol. 37, no. 8, pp. 1606–1627, Aug. 2017.

[12] J. Lee, D. Hwang, J. Park, and K.-H. Kim, "Risk Analysis and Countermeasure for Bit-Flipping Attack in LORAWAN," in 2017 International Conference on Information Networking (ICOIN), Jan 2017, pp. 549–551.

[13] A. Jacobsson, M. Boldt, and B. Carlsson, "A Risk Analysis of a Smart Home Automation System," Future Generation Computer

Systems, vol. 56, no. Supplement C, pp. 719–733, 2016. [Online]. Available: http://www.sciencedirect.com/science/article/pii/S0167739X1500281 2, [retrieved: March 2018].

[14] Y. Cherdantseva et al. "A Review of Cyber Security Risk Assessment Methods for SCADA Systems," Computers and Security, vol. 56, no. Supplement C, pp.1–27, 2016.

[15] M. Majdalawieh, F. Parisi-Presicce, and D. Wijesekera, "DNPSec: Distributed Network Protocol Version 3 (DNP3) Security Framework," Dordrecht: Springer Netherlands, 2006, pp. 227–234.

[16] TriangleMicroWorks, "Modbus and DNP3 Communication Protocols ,"2017. [Online]. Available: http://trianglemicroworks.com/docs/default-source/referenced-documents/Modbus_and_DNP_Comparison.pdf, [retrieved: March 2018].

[17] D. Wei, Y. Lu, M. Jafari, P. Skare, and K. Rohde, "An Integrated Security System of Protecting Smart Grid Against Cyber -Attacks," Innovative Smart Grid Technologies (ISGT), Gaithersburg, MD, USA, 19-21 January 2010.

[18] M. A. Rahman, and H. Mohsenian-Rad, '"False Data Injection Attacks with Incomplete Information Against Smart Power Grids," In Proc. IEEE Conf. Global Commun. (GlobeCom), Dec. 2012, pp. 3153-3158.

[19] A. Giani, et al. "Smart Grid Data Integrity Attacks," IEEE Trans. Smart Grid, vol. 4, no. 3, pp. 1244-1253, Sep. 2013.

[20] D. Li, Z. Aung, J. R. Williams, and A. Sanchez, "Efficient Authentication Scheme for Data Aggregation in Smart Grid with Fault Tolerance and Fault Diagnosis," Innovative Smart Grid Technologies (ISGT), IEEE PES, pp. 1-8, 2012.

[21] B. Vaidya, D. Makrakis, and H. Mouftah, "Secure Multipath Routing for AMI Network in Smart Grid," In Proc. IEEE 31st International Conference on Performance Computing and Communications (IPCCC), Austin, TX, pp. 408-415, 2012.

[22] Y. Yan, Y. Qian, and H. Sharif, "A Secure and Reliable In-network Collaborative Communication Scheme for Advanced Metering Infrastructure in Smart Grid," In Proc. IEEE Wireless Communications and Networking Conference (WCNC), Cancun, Quintana Roo, pp. 909-914, 2012.

[23] J. Seo and C. Lee, "The Green Defenders," IEEE Power and Energy Magazine, VOL.9, NO.1, pp. 82-90, January/February 2011.

[24] Y. Zhao, A. Goldsmith, and H. V. Poor, "Fundamental limits of Cyber Physical Security in Smart Power Grids," In Proc. 52nd IEEE Conf. Decision Control, Florence, Italy, pp. 200-205, Dec. 2013.

[25] Y. Huang, M. Esmalifalak, H. Nguyen, R. Zheng, and Z. Han, "Bad Data Injection in Smart Grid: Attack and Defense Mechanisms," IEEE Commun. Mag., vol. 51, no. 1, pp. 27-33, Jan. 2013.

[26] M. Esmalifalak, et al. "A Stealthy Attack Against Electricity Market Using Independent Component Analysis," IEEE Syst. J., to be published.

[27] G. Chaojun, P. Jirutitijaroen, and M. Motani, "Detecting False Data Injection Attacks in Ac State Estimation," IEEE Trans. Smart Grid, vol. 6, no. 5, pp. 2476-2483, Sep. 2015.

[28] F. Li, B. Luo, and P. Liu, "Secure Information Aggregation for Smart Grids Using Homomorphic Encryption," In Proc. 2010 IEEE Conf. Smart Grid Communication, pp. 327-332.

[29] X. Wang, Y. L. Yin, and H. Yu, "Finding Collisions in the Full Sha-1," In Advances in Cryptology - CRYPTO 2005: 25th Annual International Cryptology Conference, Santa Barbara, California, USA, Proceedings, 2005, pp. 17–36, Aug. 2005.

[30] R. L. Rivest, A. Shamir, and L. M. Adleman, "A Method for Obtaining Digital Signatures and Public-Key Cryptosystems," Commun. ACM, vol. 21, no. 2, pp. 120–126, 1978.

[31] N. Koblitz, "Elliptic Curve Cryptosystems," Mathematics of Computation, vol. 48, no. 177, pp. 203–209, 1987.

[32] P. W. Shor, "Algorithms For Quantum Computation: Discrete Logarithms and Factoring," In FOCS, 1994, pp. 124–134.

[33] NIST, "FIPS180-4:Secure Hash Standard," [Online]. Available: https://csrc.nist.gov/csrc/media/publications/fips/180/4/final/documents/fips180-4-draft-aug2014.pdf, [retrieved: March 2018].

[34] M. J. Dworkin, "SHA-3 Standard: Permutation- Based Hash and Extendable-Output Functions," 2015. [Online]. Available: https://www.nist.gov/publications/sha-3-standard-permutation-based-hash-and-extendable-output-functions?pub id=919061, [retrieved: March 2018].

[35] W. Mao, "Modern Cryptography: Theory and Practice," Ser. Hewlett-Packard Professional Books, Prentice Hall PTR, 2003. [Online]. Available: https://books.google.com/books?id=OOUnYAAACAAJ, [retrieved: March 2018].

[36] S. Moore, "Meet-in-the-Middle Attacks," 2010. [Online]. Available: http://stephanemoore.com/pdf/meetinthemiddle.pdf, [retrieved: March 2018].

[37] C. Adams, "Impersonation Attack. Boston, MA," Springer US, 2011, pp. 596–596. [Online]. Available: https://doi.org/10.1007/978-1-4419-5906-5 80, [retrieved: March 2018].

[38] K. Dooley, "Designing large-scale LANs," Ser. O'Reilly Series. O'Reilly,2002.[Online].Available: https://books.google.com/books?id=xwBTAAAAMAAJ, [retrieved: March 2018].

[39] J. Oh, K. Lee, and S. Moon, "How to Solve Key Escrow and Identity Revocation in Identity-Based Encryption Schemes," Berlin, Heidelberg: Springer Berlin Heidelberg, 2005, pp. 290–303. [Online]. Available: https://doi.org/10.1007/11593980 22, [retrieved: March 2018].

[40] A. Menezes, P. C. van Oorschot, and S. A. Vanstone, "Handbook of Applied Cryptography," CRC Press, 1996.

[41] Praetorian, "Man-in-the-Middle TlS Protocol Downgrade Attack,"

[42] "Linux Security Threats: The 7 Classes of Attackers," 2017. [Online]. Available: https://www.linux.com/news/chapter/Linux-security/linux-security-threats-7-classes-attackers, [retrieved: March 2018].

[43] V. Dumbrav, "Using Probability Impact Matrix in Analysis and Risk AssessmentProjects,"2013.[Online].Available: http://www.scientificpapers.org/wp-content/files/07_Dumbrava_Iacob-USING_PROBABILITY__IMPACT_MATRIX_IN__ANALYSIS_AND_RISK_ASSESSMENT_PROJECTS.pdf, [retrieved: March 2018].

[44] M. R. Bauer, "Quantum Computing is Going Commercial with the Potential to Disrupt Everything," 2017. [Online]. Available: http://www. newsweek.com/2017/04/21/quantum-computing-ibm-580751.html, [retrieved: March 2018].

[45] "NSA Suite B Cryptography - NSA/CSS." [Online]. Available: https://www.nsa.gov/ia/programs/suiteb cryptography/, [retrieved: March 2018].

[46] E. Alkim, L. Ducas, T. P¨oppelmann, and P. Schwabe, "Post-Quantum Key Exchange - a New Hope," In 25th USENIX Security Symposium, USENIX Security 16, Austin, TX, USA, August 10-12, 2016, pp.327–343. [Online]. Available: https://www.usenix.org/conference/usenixsecurity16/technical-sessions/presentation/alkim, [retrieved: March 2018].

[47] J. Hoffstein, J. Pipher, and J. H. Silverman, "NTRU: A Ring-Based Public Key Cryptosystem," In Algorithmic Number Theory, Third International Symposium, ANTS-IV, Portland, Oregon, USA, June 21-25, 1998, Proceedings, 1998, pp. 267–288. [Online]. Available: http://dx.doi.org/10.1007/BFb0054868, [retrieved: March 2018].

[48] J. M. Schanck, W. Whyte, and Z. Zhang, "Circuit-Extension Handshakes for Tor Achieving Forward Secrecy in a Quantum World," PoPETs, vol. 2016, no. 4, pp. 219–236, 2016. [Online]. Available: https://doi.org/10.1515/popets-2016-0037, [retrieved: March 2018].

[49] D. Harkins and D. Carrel, "The Internet Key Exchange (IKE)," Internet Engineering Task Force, 1998. [Online]. Available: https: //tools.ietf.org/html/rfc2409, [retrieved: March 2018].

Investigation of Technical Potentials for Load Shifting and Their Suitability to Compensate Forecast Errors of Wind Parks

Liana Maria Jacob, Sebastian Reinöhl, Wolfgang Schufft
Dept. of Power Systems and High-Voltage Engineering,
Chemnitz University of Technology
Chemnitz, Germany
e-mail: liana-maria.jacob@etit.tu-chemnitz.de
e-mail: sebastian.reinoehl@s2009.tu-chemnitz.de
e-mail: wolfgang.schufft@etit.tu-chemnitz.de

***Abstract*—Fundamental changes in the availability profile of electric power are expected due to the predominant feed-in share of renewable energy sources. In order to balance the differences between the regional generation and consumption, caused by forecast errors, the expansion of the power grid is currently being accelerated. However, a rational and resource-saving alternative constitutes the participation of consumers, which are capable of shifting load, in a Demand Side Management (DSM) concept. In the present work, the suitability of deploying existing consumers for compensating wind forecast errors in a German federal state is investigated. For the investigation, a survey regarding the structure and proportion of existing consumers in the domestic, industrial and service sector was conducted. Furthermore, the identification and quantification of their potentials of load shifting took center stage. A confrontation of expected positive, as well as negative forecast errors and technical potential of existing loads indicate that there is a high probability of successful elimination of unbalances, if a collaboration between wind parks and regional consumers was accounted.**

Keywords- Demand Side Management; forecast error; load shifting; wind parks.

I. Introduction

Future energy systems are prone to pass through a massive changeover from fossil to renewable energy generation as the German government decided to be one of the leading countries on the way to a more sustainable environment. However, renewable energy generation is often subject to uncertainties due to positive and negative forecast errors, consequently leading to high, dynamically-shaped availability profiles of the generated electric power. Since generation and consumption in the electric power system has to be kept in balance, a resource-saving manner of providing the required balancing power is indispensable. Facing these uncertainties implies assuring that online power plants dispose of sufficient generation capacity, resulting in a permanent underutilization rate of power plants. By including load shifting consumers in a Demand Side Management (DSM) concept, the operating reserve of the power plants could be minimized or even replaced, increasing their utilization rate and their efficiency.

Up to now, several scientists researched this field. Main contributions were related to the benefits and challenges of DSM [1] and to the quantification of load shifting potential of different consumers [2][3]. Technical, as well as economic aspects with respect to developing a DSM concept were analyzed [2]-[5]. In [1], Strbac points out, the process of engaging consumers in a DSM concept was especially favorable and advantageous for grids with a great share of intermittent generation sources. Nevertheless, the slow development processes were a consequence of the lack of Information and Communication Technology (ICT) infrastructure, as well as the lack of understanding of the benefits of load shifting solutions.

A vast quantification of available load shifting solutions was presented by Klobasa in [2][3]. Klobasas publications, as well as the electric grid related report [4] refer to Germany, including the domestic, industrial and service sector. Further work related to the load shifting in industrial sectors was published by Ashok et al. [6] and Paulus et al. [7]. The impact of DSM on the domestic sector was investigated by Gottwalt et al. [8] and Schlomann et al. [9].

In addition to the publications mentioned above, this paper aims not only to present and quantify the load shifting potential of the three sectors but also to confront them to common forecast errors in the wind power generation. Load shifting potential, as well as the wind power generation data series correspond to the same German federal state, presenting an individual connection between the confronted values of load shifting potentials and expectable forecast errors. More specific information about typical density functions of occurring forecast errors are presented in [10]. In [10], the author gave an overview of the magnitude of forecast errors to be expected in different regions.

Following this objective, the paper is structured as follows: In Section II, the potentials of load shifting are presented and discussed. Hereby, a differentiation between the domestic, industrial, as well as service sector was made. Section II is followed by the evaluation of expected forecast errors corresponding to the investigated federal state. Subsequently, results were discussed in Sections IV and V and afterwards, main conclusions were drawn.

II. Evaluation of Technical Potential for Load Shifting

In order to accomplish the objective of this work, it was first necessary to identify the main consumers of the federal state that have the potential of short term load shifting and then to quantify them in terms of electric power. For this purpose, a survey was conducted implying the industrial commerce chamber, the federal state statistical office, as well as relevant industrial companies in the region. These delivered important information with regard to the technical potential of load shifting in the region including parameters, such as the annual energy consumption of different loads of the domestic, industrial and service sectors in the federal state. Specific market saturation rates, as well as average electric power of appliances corresponding to the domestic sector were made available. The appliance with the highest market saturation rate among households (hh) of the investigated federal state is the refrigerator, with 99.8 %, followed by the washing machine, with 96.65 %.

In case of industrial consumers, which are able to provide load shifting potential by interrupting their production, such as the steel industry, the annual production plays a major role in determining the load shifting potentials. In case of food industry, where load shifting can be made available by cooling houses and refrigeration compressors, the technical potential of load shifting is available at each time of day. Constraints are mostly given by the admissible duration of the load shifting process. Aiming to make a distinction between air conditioning and ventilation in different branches of the industry and the cooling houses, a category named cross-section technology has been added to the calculation. This category includes the ventilation and air conditioning of production and manufacturing halls, as well as machines.

With regard to the service sector, the individual character was given by the average annual consumption as well as in most of the cases information about the surface area and the working schedules of consumers. By knowing the surface area of supermarkets, discounters or even office buildings, the demand of air conditioning, ventilation of rooms and cooling or freezing of food products can be approximated. Furthermore, water heating processes are also to be taken into account. The technical potentials of the service sector are tightly connected to the weather conditions. For this work, consumption values, which correspond to the summer months and are therefore less significant for a DSM concept were considered. If winter months are assumed, the technical potential of air conditioning in the service sector will increase.

After acquiring these information, the power to be made available for load shifting was quantified by using the methodology and Load Management Factors (LMF) thoroughly presented in [2] and [4]. Hereby, the results listed in the tables I to III with respect to the domestic, industrial and service sector, respectively, were obtained.

Table I shows the technical potentials of one single household. These amount to 146 MW if a population of 2 million households was considered. Summing up the load shifting potential of the three sectors, a value of 533.3 MW was obtained. As shown in Table I, as well as in the publications [2]-[4][8] a considerable load shifting potential of the domestic sector could be provided by appliances with great market saturation, such as the refrigerator, or the washing machine. According to [8] the load of refrigerators can only be postponed for 30 minutes. A short duration of the load shifting process reduces considerably the available shifting power that might be used in a DSM concept. Appliances, such as the washing machine or the dishwasher are connected to possible comfort limitations of the inhabitants. Therefore, the utilization of the whole calculated technical potential seems to be rather unlikely.

A survey among the industrial consumers showed that the only industrial process to be possibly used in a DSM concept can be found in the steel production. Any other industrial branch would require an absolute changeover with respect to their processes in order to be eligible for a DSM concept. Nevertheless, the activation of the technical load shifting potential of the steel industry is tightly connected to the melting process and is therefore limited.

TABLE I. Load Shifting Potentials - Domestic Sector

Application	Energy Consumption	Market Saturation	Shifting Power
Washing machine	150 kWh/a	96.65 %	23 W/ hh
Dryer	280 kWh/a	23.8 %	11 W/ hh
Dishwasher	215 kWh/a	61.2 %	19 W/ hh
Refrigerator	200 kWh/a	99.8 %	29 W/ hh
Freezer	280 kWh/a	45.1 %	17 W/ hh

TABLE II. Load Shifting Potentials - Industrial Sector

Application	Energy Consumption	LMF	Shifting Power
Steel industry	1100 GWh/a	50 %	135 MW
Food industry	155 GWh/a	50 %	17.7 MW
Cross-section technology	519 GWh/a	11 %	16.6 MW

TABLE III. Load Shifting Potentials - Service Sector

Application	Energy Consumption	LMF	Shifting Power
Cooling Processes	128 GWh/a	63 %	169 MW
Ventilation	365 GWh/a	50 %	20 MW
Heating	238 GWh/a	25 %	7 MW
Air Conditioning	259 GWh/a	75 %	22 MW

According to [2] an activation is only 40 times a year possible. A reduction of the consumption can be delayed for four hours. Cooling houses permit the load shifting up to a duration of two hours. In addition, cross-section technologies, such as cooling processes, ventilation or heating processes in the industrial sector might be available for participation in a DSM concept on daily basis with a maximum duration of two hours. The conducted survey confirmed that in the service sector only cross-section technologies can support shifting loads. A considerable amount can be provided by the consumption of cooling systems in supermarkets or discounters. The share of ventilation and air conditioning to the total technical potential is almost insignificant during the summer months.

III. EVALUATION OF EXPECTED FORECAST ERROR

The data series corresponding to the wind power generation was provided by a regional grid operator. These contain 15-minutes values describing the actual generation and the forecasted values during a year. Calculated forecast errors P_{FE} were defined as difference between the forecast generation P_F and the actual generation P_A:

$$P_F - P_A = P_{FE}, \quad (1)$$

with $P_{FE} < 0$ describing an underestimation and $P_{FE} > 0$ representing an overestimation of wind generation.

An evaluation of the given data series showed, the positive forecast error totaled up to 109 GWh/ a, while the negative forecast errors equaled -127 GWh/ a. The maximal values of the positive and negative forecasted errors were 210.58 MW and -239.66 MW, respectively, while the reference installed power of the wind turbines amounted to 1112 MW. Nevertheless, a positive median forecast error of 16.3 MW is to be assumed in case of overestimation. The median underestimation forecast error lied at -21.2 MW.

In order to get a better overview of the type and magnitude of forecast errors in the evaluated wind data series, the error magnitudes were analyzed by means of the frequency of their occurrence in the data series. The obtained histogram showed in Figure 1, indicates that great values of underestimation errors are more frequent than great values of overestimation errors. On the other hand, the majority of the forecast errors, which are positive, lie within a band of 2 MW to 6 MW. Forecast errors due to overestimation greater than 65 MW proved to be rather unusual.

For a better understanding of the character of forecast errors with respect to the magnitudes of the actual feed-in wind power, these were divided into three groups. Each of the three groups consist of 2928 elements and represent values of low, medium or high feed-in power. In accordance with the evaluated data series the low feed-in values range from 1.2 MW to 73.49 MW. The high feed-in power values lie within a band of 196.5 MW to 840 MW.

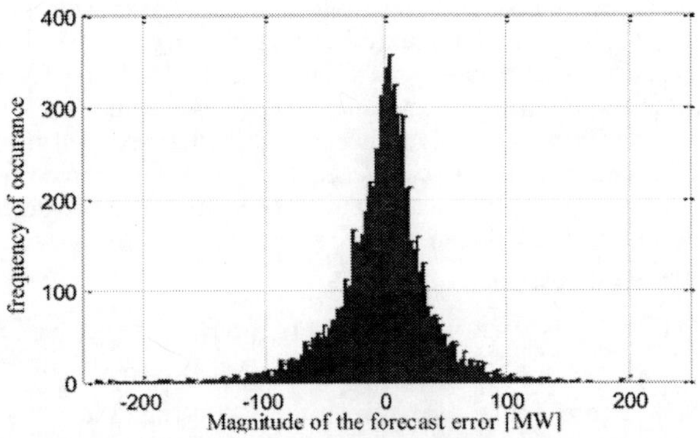

Figure 1. Relative Frequency of Forecast Errors over a year.

The tendency of overestimation or underestimation depending on the magnitude of the feed-in values was following analyzed by means of the distribution function for each of the three groups, as it can be seen in Figure 2. The average values corresponding to the three density functions differ from zero, meaning that there are certain tendencies of the three groups for overestimation or underestimation. In the area of the high feed-in power values, 196.5 MW to 840 MW, the average value of the forecast error, equal to -17.29 MW, is negative, thus preponderantly describing underestimation errors. In contrast to the high feed-in group, the density functions of the other two groups are defined by positive average values. The average value of the normal distribution of the group represented by middle feed-in power values between 73.5 MW and 196.49 MW is equal to 2.89 MW, meaning that most of the forecast errors were positive. Still, the number of overestimation and underestimation errors are almost equal for this power interval. The forecast errors of low feed-in power show a clear tendency to overestimation. Further information related to the forecast errors of the three groups, can be obtained by calculating their absolute median values. The median absolute values of the forecast errors that correspond to the three groups of high, medium and low feed-in power are 32.6 MW, 19.5 MW and 10.3 MW, respectively.

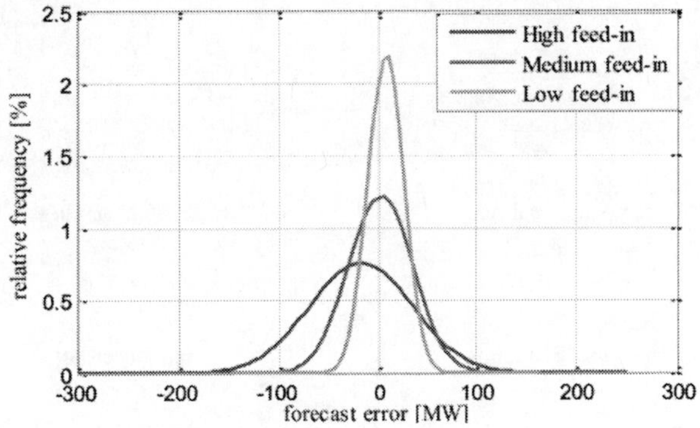

Figure 2. Normal distribution of Forecast Errors over a year of different feed-in levels.

The calculated median values support the proposition that the absolute magnitude of forecast error arises with increasing magnitudes of the actual feed-in power. Furthermore, the investigated data series showed that forecast errors are likely to occur in the afternoon and in the late evening hours, as calculated forecast errors correspond to the day-ahead forecast method. By making use of another forecast method, lower values regarding the forecast error could be achieved.

IV. Connection between Technical Potential of Load Shifting and Forecast Errors

The connection between overestimation and underestimation and the required countermeasures are described by the diagram in Figure 3. A positive forecast error resulting from overestimation would therefore be followed by a demand of positive balancing power. Positive balancing power can be provided by consumers that are able to postpone their regular consumption cycles or reduce these. This type of procedure can lead to an additional consumption at a later point of time, since most of the consumption cycles still has to be executed. Still, a short term error compensation can be achieved, if positive forecast error are balanced out by deploying the right amount of positive balancing power. In case of underestimation, negative balancing power is required. Appliances that have not been yet in operation are collectively switched on with the purpose of collecting the superfluous power available. The load that has been procured earlier may result in a reduced load profile at a later point in time. By matching the appropriate amount of negative balancing power to the occurring negative forecast error, a short term compensation of negative forecast error will result. In this way, a short term compensation of forecast errors and therefore an improvement of the electric grid by balancing supply and demand will be reached.

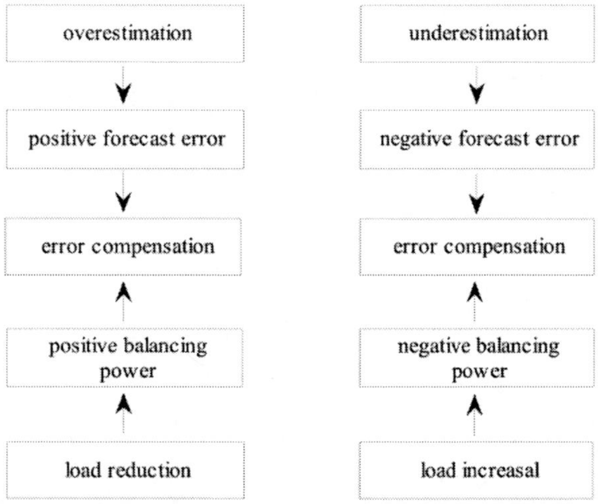

Figure 3. Connection between the type of forecast error and suitable countermeasures.

V. Interpretation of Results

In accordance to the objective of this work, a confrontation between the technical potential and the expectable forecast error in wind power generation corresponding to the federal state is indispensable. Evaluating the results on regional base aims to provide an additional share of plausibility and applicability. The assumption that an ample participation of regional consumers to a DSM concept is able to face forecast errors could be consolidated, if it was shown that technical potential for shifting loads exceeds the expectable forecast error in most of the cases. This assumption has also been considered in the work of Stötzer et al. [11]. In this work, the authors analyzed the load shifting potential in Germany and predicted the future development of the German energy system.

Developing a DSM concept requires in the first step the identification and quantification of regional technical potential. For this purpose, consumers have to be divided into two categories, providers of positive or negative balancing power. According to the regional industrial processes, an utilization of the load shifting operation is permitted only 40 days a year and is straightly oriented to steel production. For a time interval up to 4 hours, 135 MW can be made available as positive balancing power in case of overestimation. The participation of industrial consumers to compensate for positive wind forecast errors is therefore only possible to a limited extent, although to be encouraged. Cooling houses as part of the food industry could be engaged in a DSM concept based on direct or indirect control. Cooling processes of industrial sector can be activated on daily basis.

All technical potentials that are related to cross-section technologies, such as cooling processes, ventilation or air conditioning can be made available every day of a year at least once a day for a time interval up to two hours. These can be aggregated and then utilized in order to provide positive, as well as negative balancing power, offering a secure source of technical potential for load shifting. As already stated in the work of Stötzer et al. [11], these are associated to a very low limitation of comfort for each consumer in different sectors due to their inherent storage function.

Technical potential of the domestic sector resulting from appliances, such as washing machines or dishwashers are mostly available during the day. Additionally, they imply the active participation of the inhabitants to the DSM concept. Given the high market saturation of this appliances, it might be realistic that part of the available households engage in a DSM concept. Every source of load shifting potential, which can be made available on daily basis, is to be approximated by taking into consideration standardized load profiles of the consumers mentioned in the tables 1 to 3. The reference load profiles describing appliances of domestic sector were developed by Stadler [12]. Other load profiles can be found in [4]. Results representing the profile of the total amount of available shifting load are shown in Figure 4. It has to be mentioned that the presented values represent the maximum available amount of load shifting potentials. These are only once available during a day. If potentials have already been called up, these will no longer be available for a certain period of time and are to be subtracted from the given maximum amount. By comparing the values related to the maximum expected positive or negative forecast error, 210 MW and - 239 MW, respectively, and the maximum amount of load shifting power that could be provided by regional consumers, it can be easily recognized that the technical available load shifting potential can compensate expected errors in most of the cases. The maximum negative forecast error, with an expected magnitude equal to - 239 MW exceed the technical potential available during the night. However, forecast errors of similar magnitudes are rather unusual and are mostly to be attributed to very particular events, such as accidents or storms that put entire wind parks out of service. Furthermore, forecast errors due to overestimation lie in over 90 % of the cases within a band of 0 and 25 MW. Over 85 % of the underestimation errors do not exceed 25 MW. Taking into consideration the presented profile in Figure 4, a forecast error of this extent could be easily compensated by making use of regional load shifting potentials. Even if forecast errors up to 25 MW occurred simultaneously or repeatedly, there would be a reasonable chance that the required share of load shifting potential was available. Due to the fact that great forecast errors are more likely to occur during the late evening hours, the chance to compensate them by means of load shifting, increases.

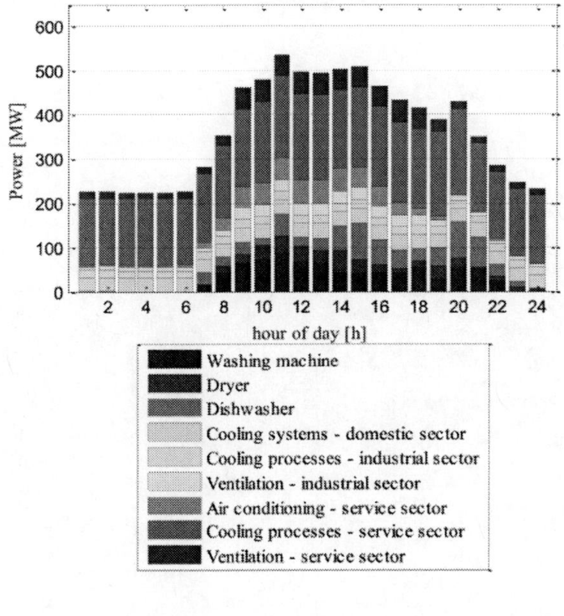

Figure 4. Profile of Available Load Shifting Power

VI. Conclusion and Future Work

Based on a survey, an investigation of the technical potential of a German federal state was performed. The results are in accordance with already published results regarding Germany. These showed that every sector can be part of a DSM concept and contribute to it. A total technical potential equal to 533.3 MW was calculated. This technical potential could be activated only once during a day, assuming that all participants are able to shift load. During the night, over 200 MW technical potential for load shifting can be made available.

Most reasonable participants to a DSM concept turned out to be the cross-section technologies in the service and industrial sector, as these can be activated at least once a day. A further option is represented by appliances of the food industry and the food retail like cooling houses or supermarkets with high surface areas. These have a high power consumption and a forecast error could be compensated by including only a small number of participants. Technical potential for load shifting connected to industrial production processes is only represented by the steel industry. To a limited extent, a load reduction can be achieved. However, there is a huge need of modernization in this sector so that processes can be deployed for this purpose.

The domestic sector also shows a high technical potential for load shifting. A direct load control could be applied to cooling systems. Still, it has to be highlighted that the duration of this process shouldn't exceed one hour. Further appliances in the domestic sector to contribute to a DSM concept are washing machines or dishwashers, which might be connected to comfort limitations of the inhabitants. It also has to be mentioned that including diverse appliances related to the domestic sector in a DSM concept is associated to an elaborate control concept and vast communication network.

A reduction of the ICT infrastructure development and though a relief of the transition to a smart grid can be achieved, if every wind park had a few regional partners, which are willing to deploy their technical potential of load shifting, if required by the wind park or grid operators. Nevertheless, deploying available technical potential in order to compensate forecast errors is only possible if all involved participants are likely to profit from it. The economic benefits of their participation should definitely exceed the resulting effort or the comfort limitations.

Future research should consider the effects of other forecast methods on the resulting forecast error magnitudes in more detail. In addition, the development of an algorithm that identifies and quantifies load shifting potential, as well as the magnitude of the forecast error to be expected in a given electric grid is desirable in order to prove that there is a high probability of successful elimination of unbalances in diverse grids.

REFERENCES

[1] G. Strbac, "Demand side management - Benefits and challenges," Energy Policy, vol. 36, pp. 4419–4426, December 2008, doi: 10.1016/j.enpol.2008.09.030.

[2] M. Klobasa, Dynamic simulation of a load management and integration of wind energy into a national electricity grid with regard to control cost aspects. (in german) Doctoral Thesis, Zürich, 2007. Available from:

http://publica.fraunhofer.de/dokumente/N-68615.html, retrieved: 03.2018.

[3] M. Klobasa, "Analysis of demand response and wind integration in Germany's electricity market," IET Renewable Power Generation, vol. 4, pp. 55–63, January 2010, doi: 10.1049/iet-rpg.2008.0086.

[4] S. Kohler, A. Agricola, and H. Seidl: Integration of renewable energies into the German electricity supply in the period 2015-2020 with outlook 2025 (in german). Berlin, 2010. Available from:

https://www.dena.de/fileadmin/user_upload/Download/Dokumente/Studien___Umfragen/Endbericht_dena-Netzstudie_II.PDF, retrieved: 03.2018.

[5] E. Birrer, D. Bolliger, R. Kyburz, A. Klapproth, and S. Summermatter, "Load Shift Potential Analysis Using Various Demand Response Tariff Models on Swiss Service Sector Buildings," Presented at the 8th international conference on energy efficiency in domestic appliances and lighting—EEDAL'15, Lucerne.

[6] S. Ashok and R. Banerjee, "Load-management applications for the industrial sector," Applied Energy, vol. 66, pp. 105–111, June 2000, doi: 10.1016/S0306-2619(99)00125-7.

[7] M. Paulus and F. Borggrefe, "The potential of demand-side management in energy-intensive industries for electricity markets in Germany," Applied Energy, vol. 88, pp. 432–441, February 2011, doi: 10.1016/j.apenergy.2010.03.017.

[8] S. Gottwalt, W. Ketter, C. Block, J. Collins, and C. Weinhardt, "Demand side management-A simulation of household behavior under variable prices," Energy Policy, vol. 39, pp. 8163–8174, December 2011, doi: 10.1016/j.enpol.2011.10.016.

[9] B. Schlomann et al., Energy consumption of the trading, commercial and service sectors in Germany from 2007 to 2010 (in german). Available from:

http://docplayer.org/2918561-Energieverbrauch-des-sektors-gewerbe-handel-dienstleistungen-ghd-in-deutschland-fuer-die-jahre-2007-bis-2010.html, retrieved: 03.2018.

[10] B.-M. Hodge et. al., "Wind Power Forecasting Error Distributions: An International Comparison," Tech. Rep., 2012.

[11] M. Stötzer, I. Hauer, M. Richter, and Z.-A. Styczynski, "The potential of demand-side management in energy-intensive industries for electricity markets in Germany," Applied Energy, vol. 146, pp. 344–352, May 2015, doi: 10.1016/j.apenergy.2015.02.015.

[12] I. Stadler, Demand Response – Non-electric storage systems for electricity supply systems with a high share of renewable energies (in german). Doctoral Thesis , Berlin, 2005.

QoS-compliant Data Aggregation for Smart Grids

Jad Nassar
HEI - Yncrea HdF, France
Inria, France
Email: jad.nassar@{yncrea,inria}.fr

Nicolas Gouvy
HEI - Yncrea HdF, France
Email: Nicolas.gouvy@yncrea.fr

Nathalie Mitton
Inria, France
Email: Nathalie.mitton@inria.fr

***Abstract*—The Smart Grid (SG) aims to transform the current electric grid into a "smarter" network where the integration of renewable energy resources, energy efficiency and fault tolerance are the main benefits. A Wireless Sensor Network (WSN) controlling and exchanging messages across the grid is a promising solution because of its infrastructure free and ease of deployment characteristics. This comes at the cost of resource constrained and unstable links for such networks. The management of communication is then an issue: billions of messages with different sizes and priorities are sent across the network. Data aggregation is a potential solution to reduce loads on the communication links, thus achieving a better utilization of the wireless channel and reducing energy consumption. On the other hand, SG applications require different Quality of Service (QoS) priorities. Delays caused by data aggregation must then be controlled in order to achieve a proper communication. In this paper, we propose a work in progress, that consists of a QoS efficient data aggregation algorithm with two aggregation functions for the different traffics in a SG network. We expect to reduce the energy consumption while respecting the data delivery delays for the different SG applications.**

Keywords–Smart Grid; Data Aggregation; QoS; Wireless Sensor Networks

I. INTRODUCTION

Wireless Sensor Networks (WSNs) are a potential candidate in the transition from the actual grid to the Smart Grid (SG), where the integration of renewable energy resources, energy efficiency and fault tolerance are the main benefits. On the other hand, WSNs are resource restrained entities with limited computing capabilities, even if in a SG electricity and energy exists, but connecting sensors to such high voltage with intermittent and ill-adapted energy levels is sometimes inappropriate. For that, battery-powered sensors must be deployed all over the grid alongside with the main-powered ones. This will raise a challenge in the data collection process, specially in a SG network where billions of packets with different sizes and priorities are frequently sent within the network. Data aggregation is a feasible paradigm that consists of combining data from multiple sensors across the network and sending the aggregated data to the base station. This will reduce loads on the communication links, thus achieving a better utilization of the wireless channel and reducing energy consumption. In a SG, different applications require different Quality of Service (QoS) priorities. Consequently, data aggregation must respect these requirements (i.e, delays caused by aggregating the packets) in order to ensure a proper communication. Therefore, in this paper, we propose a QoS efficient data aggregation algorithm for the different traffic in a SG network. The rest of the paper is organized as follows: Section II presents prior work on data aggregation in WSN. Section III describes our proposed solution. Section IV discusses some relevant issues about our proposal and expected results. Finally, section V concludes the paper.

II. RELATED WORKS

Many works addressed the data aggregation in WSNs and SGs. In [1], two aggregation methods for processing data in smart meters were used: combining and manipulating. In the combining method, the concentrator removes all individual headers and includes only one single header for the large packet with no data modifications. The manipulating method consists of calculating the result of the messages thus reducing considerably the total size of the messages. Data packet concatenation in SGs was also addressed in [2], they achieved header compression on packets and formulated an optimization problem to optimally configure the sizes of the aggregated packets. However, they considered only overhead reduction, which may be insufficient alone in the presence of bigger data packets with smaller headers. Many other researches considered energy [3], delay guarantee [4] and other QoS requirements [5] [6] in data aggregation for SGs and sensor networks generally. However, none of these works addressed the challenge of having delay sensitive data traffic with different delivery priorities and sizes while reducing energy consumption and maximizing the available bandwidth.

III. PROPOSED SOLUTION

In our proposition, we consider a SG network consisting of several wireless sensors collecting data with different packet sizes and priorities. They can potentially act as aggregators, if they have enough resources, that receive the data and aggregate or concatenate it depending on their QoS requirements, and finally send the aggregated data across the network. We note that the routing process is mostly left unchanged, we only add the aggregation functionality when it is possible. Packets are generated with classifiers in their headers considering their type and criticality, we classify them into two levels: critical and regular. We note that these levels could be adjusted for other applications depending on the network characteristics. Two different queues are created on the aggregator level: Lossy and Lossless queue. The lossy queue contains delay insensitive data packets (regular) that are generally big in size [7], which will allow us to aggregate the packets with the appropriate aggregation function [8]. The lossless queue contains delay sensitive data packets (critical) with critical priorities and with a header which represents a significant overhead compared to the payload size. Header compression is thus performed on the packets.

Algorithm 1: Aggregator node

```
Update_AD() ;      // AD= Maximum allowed delay - time
  from node to aggregator
if Free_Space() AND Battery_Node > Threshold1 then
  | Aggregate();
  | Send_AgPacket();
else
  | Send_Packet();
end
```

Figure 1. Aggregator node algorithm

Algorithm 2: Lossy_Queue()

```
Init: TTL
if Earliest_Deadline > Delivery_Threshold AND AD > Delivery_Threshold
  AND AggrPktSize < MTU AND TTL > 0 then
  | Lossy_Aggregation();
  | Update_Earliest_Deadline() ; // Earliest deadline in the
  |   aggregated packet
  | TTL - - ;
else
  | Concatenate();
end
```

Figure 2. Lossy queue algorithm

The proposed aggregation algorithm

In the following, we will explain the main functionality of the aggregation algorithm and its functions. The sender node sends and receives packets from other sensor nodes with different priorities included in their headers. Routes are constructed according to the existing routing protocol with no influence for the aggregation algorithm as already mentioned. In figure 1 when an aggregator receives a packet it will firstly update its delivery time (*Update_AD()*) corresponding to the timestamp included in the header of the packet (maximum allowed delay) minus the time the packet spent to arrive to the aggregator. This will allow us to identify how much time the packet can stay in the aggregator. We store the value in the variable *AD*, we call it maximum allowed delay left. After that, the function *Free_Space()* will check whether the node can store more packets, and the function *Battery_Node()* will check whether the node has enough energy (more than a predefined threshold) to aggregate more packets. If these two conditions hold, we can aggregate packets and send the aggregated packets afterwards. If not, the packets are sent without aggregation. In the aggregate function, we check the *Pkt_type* and send it to the corresponding queue. If the packet is tagged regular, it is sent to the *Lossy_Queue()* (figure 2), where four conditions have to be validated in order to aggregate packets:

- *Earliest_Deadline > Delivery_Threshold*: aggregating if the packet with the earliest deadline in the aggregated packet is still within its allowed delay.
- *AD > Delivery_Threshold*: which means that the Delivery threshold from the aggregator to the sink must be smaller that the Maximum allowed delay left.
- *AggrPktSize < MTU*: aggregating as long the aggregated packet is smaller than the Maximum Transmission Unit (MTU) of the link.
- *TTL > 0*: If the above conditions were not valid yet and after a certain time we send the packet anyway on the link.

As long as these above conditions are valid, an arriving regular packet to the aggregator will undergo a *lossy_aggregation()*, and the timers are updated. Same applies for the *Lossless_Queue()* with a packet tagged critical. If not valid, we concatenate the incoming packet with the existing aggregated packet if possible and send it immediately. We note that the sink node will send back with the acknowledgment the time the packet spent from the aggregator to the sink in order to update the *delivery threshold*.

IV. Discussion and Expected Results

First of all aggregating packets will lead to less packets sent across the network and less bandwidth consumed, which will result in reducing the load on the communication links and achieving energy savings since the communication task consumes most of the energy in WSNs. Moreover, packets criticality and sizes are taken into consideration in our proposition. For that we expect that the packets will arrive within their deadlines, thanks to the different timers and thresholds across the network. We note that the delivery delays will be longer than a non aggregation scenario where packets are not stored in the queues. Packet delivery ratio might be affected also in our proposition, since aggregating means sending bigger packets thus resulting in more losses. We will deeply investigate this issue in order to mitigate these potential losses. Moreover, we will investigate in future works the impact of disaggregation at the destination node.

V. Conclusion

In this paper, we propose a work in progress solution for data aggregation in SGs networks. QoS requirements of the different applications are taken into consideration by storing the packets in two different queues depending on their quality requirements. The expected results will reduce the energy consumption in a SG controlled by a WSN, while respecting the corresponding delays and QoS requirements. Several tests and investigations have to be performed (i.e, computer simulations) before the completion of this work, after that we will test our algorithm on a real test bed [9] to validate our theoretical approach.

Acknowledgment

This work was partially funded by a grant from the MEL (Métropole Européenne de Lille), the SoMel SoConnected project (Ademe, PIA2), Yncréa Haut-de-France and CPER Data.

References

[1] T. Shiobara, P. Palensky, and H. Nishi, "Effective metering data aggregation for smart grid communication infrastructure," in 41st Annual Conference of the Industrial Electronics Society (IECON). IEEE, 2015, pp. 002 136–002 141.

[2] B. Karimi, V. Namboodiri, and M. Jadliwala, "Scalable meter data collection in smart grids through message concatenation," IEEE Transactions on Smart Grid, vol. 6, no. 4, 2015, pp. 1697–1706.

[3] F. Uddin, "Energy-aware optimal data aggregation in smart grid wireless communication networks," IEEE Transactions on Green Communications and Networking, vol. 1, no. 3, 2017, pp. 358–371.

[4] T.-C. Lee and Z. Tsai, "On the capacity of smart grid wireless backhaul with delay guarantee and packet concatenation," IEEE Systems Journal, pp. 2628–2639.

[5] P. Teymoori, M. Kargahi, and N. Yazdani, "A real-time data aggregation method for fault-tolerant wireless sensor networks," in Proceedings of the 27th Annual Symposium on Applied Computing. ACM, 2012, pp. 605–612.

[6] T. Abdelzaher, T. He, and J. Stankovic, "Feedback control of data aggregation in sensor networks," in 43rd Conference on Decision and Control. CDC, vol. 2. IEEE, 2004, pp. 1490–1495.

[7] N. Cam-Winget, J. Hui, and D. Popa, "Applicability statement for the routing protocol for low power and lossy networks (rpl) in ami networks," Working Draft, IETF, Internet-Draft. draft-ietf-roll-applicability-ami-13, April 2016.

[8] M. Grabisch, J.-L. Marichal, R. Mesiar, and E. Pap, "Aggregation functions: means," Information Sciences, vol. 181, no. 1, 2011, pp. 1–22.

[9] C. Adjih, E. Baccelli, E. Fleury, G. Harter, N. Mitton, T. Noel, R. Pissard-Gibollet, F. Saint-Marcel, G. Schreiner, J. Vandaele et al., "Fit iot-lab: A large scale open experimental iot testbed," in Internet of Things (WF-IoT), 2015 IEEE 2nd World Forum on. IEEE, 2015, pp. 459–464.

Performance Prediction of Geophysics Numerical Kernels on Accelerator Architectures

Víctor Martínez, Matheus S. Serpa, Philippe O. A. Navaux
Informatics Institute
UFRGS, Brazil
email: {victor.martinez, msserpa, navaux}@inf.ufrgs.br

Edson L. Padoin
Department of Exact Sciences and Engineering
UNIJUI, Brazil
email: padoin@unijui.edu.br

Jairo Panetta
Computer Science Division
ITA, Brazil
email: jairo.panetta@gmail.com

Abstract—**In order to develop geophysics tools for exploration of energetic resources, numerical models are proposed to understand complex geological structures. They are solved from the discretization of Partial Differential Equations by the Finite Differences Method. This method creates a pattern that solves each point in a 3D domain, and it replicates the same calculation to compute all the data domain. Because of the quantity of calculations, solving the numerical kernels requires High Performance Computing. However, the complexity of current architectures may reduce the efficiency. In some cases, applications tuning have been used to improve the performance. In this context, predicting the performance from input parameters is a critical problem. This is particularly true regarding the high number of parameters to be tuned both at the hardware and the software levels (architectural features, compiler flags, memory policies, multithreading strategies). This work focuses on the use of Machine Learning to predict the performance of geophysics numerical kernels on many-core architectures. Measures of hardware counters on a limited number of executions are used to build our predictive model. We have considered three different kernels (7-point Jacobi, seismic and acoustic wave propagation) to demonstrate the effectiveness of our approach. Results show that the performance can be predicted with high accuracy.**

Keywords–Machine Learning; Geophysics Applications; Many-core Systems; Performance Model

I. INTRODUCTION

Geophysics modeling remains fundamental to keep up with the demand for energetic resources. Thus, Oil and Gas industries rely on High Performance Computing (HPC) software as an economically viable way to reduce risks. Wave propagation algorithms are routinely used both in the oil and gas industry and in strong motion analysis in seismology. The finite-differences numerical method used for this problem also lies at the heart of a significant fraction of numerical solvers in other fields. In terms of computational efficiency, one of the main difficulties is to deal with the disadvantageous ratio between the limited point-wise computations and the intensive memory access required, leading to a memory-bound situation [1].

The objective of HPC applications is to optimize the performance. This comes from the increasing of complexity for many interdependent factors: vectorization, compiler optimizations, non-uniform memory access and several levels of memory, etc. Although a large body of literature on the optimization of this class of applications is available, predicting the performance on current architectures remains a challenge and it requires to search in a large set of input configurations [2]. On the other hand, Machine Learning (ML) is a comprehensive methodology for optimization that could be applied to find patterns on a large set of parameters. Recently, in [3] the authors present a methodology based on ML algorithms and simulation to obtain dynamic scheduling policies, whereas in [4] the authors proposed an ML-based scheme to tune the storage system and increase the I/O throughput.

This research was developed in the context of High Perfomance Computing for Enery Project (HPC4E). In this paper, we describe the procedure to predict the performance of stencil applications by a ML-based model, on accelerators architectures. The paper is organized as follows: Section II provides the fundamentals of geophysics models under study; Section III describes the methodology of our ML-based approach; Section IV presents the experiments, the performance prediction, and the model accuracy; Section V describes the related work. Finally, Section VI concludes this paper.

II. NUMERICAL KERNELS

In this section, we present the geophisycs numerical models. Due to its simplicity, the Finite-Differences Method (FDM) is widely used to design the geophysics models, when discretizing Partial Differential Equations (PDE). From the numerical analysis point of view, the FDM computational procedure consists in using the neighboring points in the north-south, east-west and forward-backward directions to evaluate the current grid point in the case of a three-dimensional Cartesian Grid. The algorithm then moves to the next point applying the same computation until the entire spatial grid has been traversed. The number of points used in each direction depends on the order of the approximation. This procedure is called stencil-based computation. The stencil sweep can be expressed as a triply nested loop presented in Figure 1.

```
for each timestep do
    compute in parallel
    for each block in X-direction do
        for each block in Y-direction do
            for each block in Z-direction do
                compute stencil(3D tile)
            end for
        end for
    end for
end for
```

Figure 1. Pseudocode for stencil algorithms.

A. 7-point Jacobi

The stencil model of 7-point Jacobi is given by the explicit 3D heat equation described in [5] and presented in (1):

$$\begin{aligned} B_{i,j,k} =& \alpha A_{i,j,k} \\ &+ \beta(A_{i-1,j,k} + A_{i,j-1,k} + A_{i,j,k-1} \\ &+ A_{i+1,j,k} + A_{i,j+1,k} + A_{i,j,k+1}) \end{aligned} \quad (1)$$

This stencil performs a single Jacobi (out-of-place) iteration. Thus, reads and writes occur in two distinct arrays (A, B), where each subscript represent the 3D index into array A or B. For each grid point, this stencil will execute 8 floating point operations [6]. Figure 2 illustrates the size of Jacobi stencil.

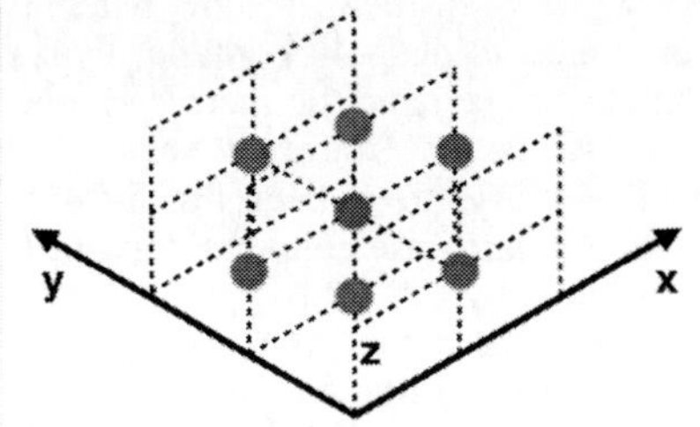

Figure 2. Seven-point Jacobi stencil [7].

B. Seismic Wave Propagation

The seismic waves radiated from an earthquake are often simulated under the assumption of an elastic medium although the waves attenuate due to some anelasticity. This numerical kernel corresponds to the discretization of the elastodynamics equation and it is of great importance both for seismic hazard assessment, as well as for the oil and gas industry. In our case, we consider a standard fourth order in space and second order in time approximation. This algorithm corresponds to the evaluation of six stress components (three in the diagonal direction and three off-diagonal) and three velocity components. A detailed description of the numerical modeling of seismic waves on multi-core platforms is presented in [8].

$$\begin{aligned} \sigma_{xx}^{n+1}(i,j,k) =& \sigma_{xx}^{n}(i,j,k) \\ &+ A_1[a_1(V_x^n(i+\frac{1}{2},j,k) - V_x^n(i-\frac{1}{2},j,k)) \\ &+ a_2(V_y^n(i,j+\frac{1}{2},k) - V_y^n(i,j-\frac{1}{2},k)) \\ &+ a_3(V_z^n(i,j,k+\frac{1}{2}) - V_z^n(i,j,k-\frac{1}{2}))] \\ &+ B_1[a_1(V_x^n(i+\frac{3}{2},j,k) - V_x^n(i-\frac{3}{2},j,k)) \\ &+ a_2(V_y^n(i,j+\frac{3}{2},k) - V_y^n(i,j-\frac{3}{2},k)) \\ &+ a_3(V_z^n(i,j,k+\frac{3}{2}) - V_z^n(i,j,k-\frac{3}{2}))] \end{aligned} \quad (2)$$

Equation (2) provides a synthetic view of the computation of one of the diagonal components, where i, j, k represent a tensor field component in Cartesian coordinates (x, y, z), and V and σ represent the velocity and stress fields, respectively. Figure 3 illustrates the size of the seismic wave propagation stencil applied to calculate velocity and stress components.

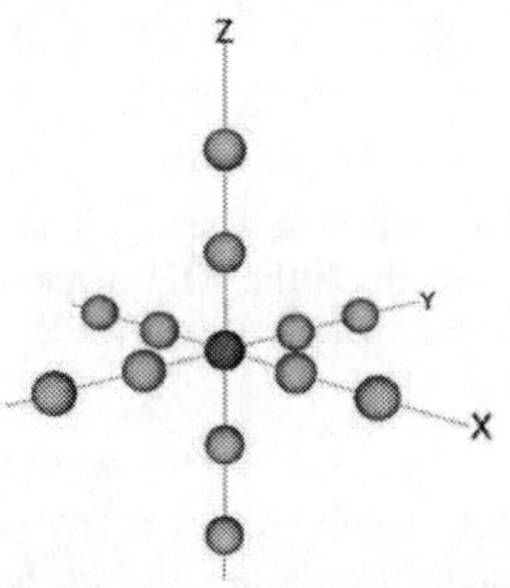

Figure 3. Seismic wave propagation stencil [9].

C. Acoustic Wave Propagation

The acoustic wave propagation approximation is the current backbone for seismic imaging tools. It has been extensively applied for imaging potential oil and gas reservoirs beneath salt domes. We consider the model formulated by the isotropic acoustic wave propagation under Dirichlet boundary conditions over a finite 3D rectangular domain, prescribing to all boundaries, and the isotropic acoustic wave propagation. Numerical method solves (3) and is detailed in [10].

$$\begin{aligned} C_{i,jk} =& a_0 C_{i,j,k} \\ &+ a_1(C_{i-1,j,k} + C_{i+1,j,k} + C_{i,j-1,k} \\ &+ C_{i,j+1,k} + C_{i,j,k-1} + C_{i,j,k+1}) \\ &+ a_2(C_{i-2,j,k} + C_{i+2,j,k} + C_{i,j-2,k} \\ &+ C_{i,j+2,k}, + C_{i,j,k-2} + C_{i,j,k+2}) \\ &+ a_3(C_{i-3,j,k} + C_{i+3,j,k} + C_{i,j-3,k} \\ &+ C_{i,j+3,k} + C_{i,j,k-3} + C_{i,j,k+3}) \end{aligned} \quad (3)$$

Propagation speed depends on variable density, the acoustic pressure, and the media density. Figure 4 illustrates the size of the acoustic wave propagation stencil.

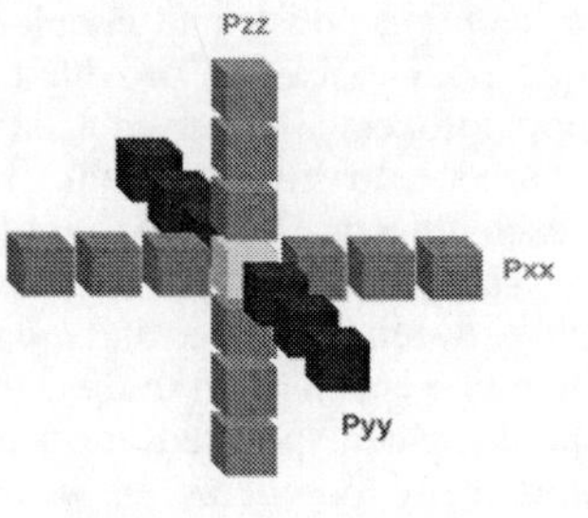

Figure 4. Acoustic wave propagation stencil [10].

Petrobras, the leading Brazilian oil company, provides a standalone mini-app of the previously described numerical method. The code was written in standard C and leverage from OpenMP directives for shared-memory parallelism. But Indeed, the parallelization strategy relies on the decomposition of the three-dimensional domain based on OpenMP loop features.

III. TESTBED AND PREDICTION MODEL METHODOLOGY

In this section, we describe our testbed, the runtime configurations and introduce the ML model.

A. Testbed

We used the Intel Xeon Phi many-core platform to carry out the experiments (code-name *Knights Landing, or KNL*). The detailed configurations are shown in Table I.

TABLE I. DESCRIPTION OF THE INTEL XEON PHI (KNL) ARCHITECTURE.

Processor	Intel Xeon Phi 7520
Clock(GHz)	1.40
Cores	68
Sockets	1
Threads	272
L2 cache size (MB)	34

The KNL brings in new memory technology, a high bandwidth on package memory called Multi-Channel DRAM (MCDRAM). MCDRAM is a high bandwidth, low capacity (up to 16GB) memory. MCDRAM has three memory modes and can be configured as *cache* mode to work as a third level cache; in *flat* mode, both the MCDRAM memory and the DDR memory act as regular memory and are mapped into the same system address space as a distinct NUMA node (allocatable memory); and the *hybrid* mode, the MCDRAM is partitioned such that either a half or a quarter of the MCDRAM is used as cache, and the rest is used as flat memory [11] [12].

The scientific applications implemented on the KNL architectures are developed in openMP, and as we can explained in Section II a stencil is built by three nested for loops. Then, scheduling influences the application performance. OpenMP scheduling can be defined in runtime by the `OMP_SCHEDULE` environment variable. This variable is a string formatted by two parameters: scheduling policy and chunk size. Four different loop scheduling policies can be provided to OpenMP: *Static* divide the loop into equal-sized chunks; *Dynamic* uses the internal work queue to give a chunk-sized block of loop iterations to each thread; *Guided* is similar to dynamic, but the chunk size starts off large and decreases to better handle load imbalance between iterations; and *auto*, when the decision regarding scheduling is delegated to the compiler. The optional parameter (chunk), when specified, must be a positive integer and defines how many loop iterations will be assigned to each thread at a time [13].

Based on this platform, Table II details all the available configurations for the optimization categories. As it can be noted, a brute force approach would be unfeasible due to the large number of simulations required (443,904), because some of these executions can take many hours (or days).

TABLE II. AVAILABLE CONFIGURATIONS FOR THE OPTIMIZATION PROCEDURE.

Optimization	Parameters	Total configurations
Number of threads	1	272
Chunk size	1	272
Scheduling policy	1	3
Memory mode	1	2
Total	**3**	**443,904**

B. Feature vectors

We emphasize that selection of the relevant feature vectors is a key ingredient of our method, which are described below:

1) The **Runtime vector** is defined by OpenMP implementation features such as the number of threads (`OMP_NUM_THREADS` environment varible), the loop scheduling policy (static, dynamic or guided), the chunk size, and the memory mode (cache or flat).
2) The **Hardware Counters vector** is built on top of PAPI library, to collect the most relevant metrics from the hardware counters: total of L2 cache misses (measured by `PAPI_L2_TCM` event), and total of cycles (measured by `PAPI_TOT_CYC` event) [14]. We decided to use related cache values because stencils are memory bounded problems, and the number of available counters is determined by the architecture.
3) The **Performance vector** represents the total elapsed time to solve the geophysical problem.

As we can see, the performance depends on several parameters that create a n-dimensional problem and if we try to model it by a regression method it can not be solved by 2D or 3D classical models.

C. Prediction Model

The proposed ML model is based on Support Vector Machines (SVM), which is a supervised ML approach introduced in [15] and extended to regression problems where support vectors are represented by kernel functions [16]. The main idea of SVMs is to expand hyperplanes through the output vector. It has been employed to classify non-linear problems with non-separable training data by a linear decision surface (i.e., hardware counters behavior in previous Section III-B).

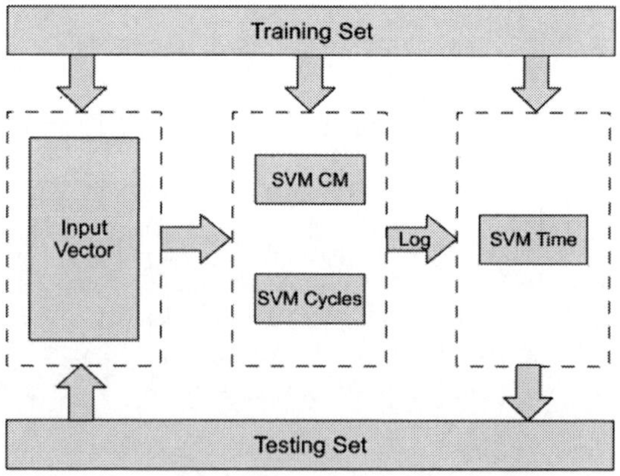

Figure 5. Flowchart of the proposed ML-based model.

In our case, we considered a classical ML model with three layers of measurements (input, hidden, and output) related with vectors explained in Section III-B, and extends the work done by [17], (Figure 5). The input layer contains the values from the runtime vector. The hidden layer contains two SVMs that take values from the input vector to simulate the behavior of hardware counters. Finally, the output layer contains one SVM that takes each simulated value from the hidden layer to obtain the predicted execution time. Our model was implemented by *e1071* R package.

IV. Experimental results

In this section, we present the results of our prediction model. We used a three-dimensional grid of size 512x512x512, and 190 time iterations, as benchmark for our experiments.

A. Training and validation sets

We created a training set by randomly selecting a subset from the configuration parameters presented in Table II. Then, for each experiment we measured the hardware counters (L2 cache misses, and total cycles) and the performance (execution time). Because hardware counters have very large values, it was necessary to perform a dynamic range compression based on a logarithmic transformation, between the hidden layer and the output layer, as shown in Figure 5.

A random testing set was used since all SVMs in both the hidden and the output layers are trained to predict the execution time values. After that, we measured the accuracy of the model using statistical estimators. Table III presents the total number of experiments that were performed to obtain the training and validation sets. It is remarkable that total number of experiments used for testing and validation, for each stencil, is lower than 1% of total configurations described in Table II.

TABLE III. NUMBER OF EXPERIMENTS

	Training	Testing	**Total**
Jacobi	334	3007	**3341**
Seismic	346	3122	**3468**
Acoustic	335	3021	**3356**

B. Hardware Counters Behavior

Figure 6 illustrates how the hardware counters measurements are affected by the input variables. Each point represents one experiment when varying the input parameters described in Section III-B; the X domain represents the L2 cache misses, the Y axis represents the total cycles, and the color represents the different values for the input parameter. For instance, Figure 6(a) represents the scheduling policy (green is dynamic, red is guided, and blue is static) for the Jacobi stencil. We can see how the static scheduling tends to be separated from other values. Figure 6(b) shows the number of threads used to solve the seismic stencil. We also can see how each value create one easily separated area. Figure 6(c) also presents how chunk size trends to creates separated areas for the acoustic stencil. We can resume this behavior as follows: the input values changing affects the application performance and creates several separated areas in the graphic representation. As each color represents a different value for the input value these areas could be separated by hyperplanes from a SVM.

C. Prediction Results

We evaluate the model with two statistical estimators: the Root Mean Square Error (RMSE) and the Coefficient of Determination (R-squared); they represent the standard deviation of prediction errors and how close the regression approximates the predicted and actual data, respectively. As it can be noted in Table IV, the regression model is highly accurate. For the 7-point Jacobi implementation, the model presented an accuracy of up to 97.39%. For the seismic wave implementation, on the other hand, the model presented an accuracy of up to 85.62% and the acoustic wave propagation model has 93.2% accuracy. These results are similar to prediction of multi-core architectures presented in [18] and [17].

TABLE IV. STATISTICAL ESTIMATORS OF OUR PREDICTION MODEL

	RMSE	R-squared
Jacobi	2.9894	0.9739
Seismic	21.0183	0.8562
Acoustic	40.0748	0.9322

V. Related Work

Recent HPC architectures, including accelerators and co-processors, proved to be well suited for geophysics simulations, outperforming the general purpose processors in efficiency. And some works are developed to optimize and to predict the performance. Because numerical kernels are a memory-bound problem, common optimizations are cache-based. In [19], the authors worked on target cache reuse methodologies, and demonstrated that memory system organization reduce the efficacy of traditional cache-blocking optimizations for stencil computation on multiple architectures (multi-core and accelerators). In the same way, in [8], the authors worked on target cache reuse methodologies across single and multiple stencil sweeps, examining cache-aware algorithms as well as cache-oblivious techniques in order to build robust implementations.

At high level, adjusting the runtime parameters can optimize the performance of stencil applications. In [20], the authors focused on acoustic wave propagation equations, choosing the optimization techniques from systematically tuning the algorithm. The usage of collaborative thread blocking, cache blocking, register re-use, vectorization and loop redistribution. In [2], the authors analyzed the performance of task scheduling algorithms. They concluded that different scheduling policies combined with different task sizes may affect the efficiency and performance of seismic wave kernels. In [21], the authors studied the effect of different optimizations on elastic wave propagation equations, achieving more than an order of magnitude of improvement compared with the basic OpenMP parallel version.

Methods for automatic code generation are used at runtime level. In [22], the authors presented a stencil auto-tuning framework for multi-core architectures that converts a sequential stencil expression into tuned parallel implementations. In [23], the authors present an automatic source-to-source transformations framework. Thus, in [24], the authors suggest using runtime reconfiguration, and a performance model, to reduce resource consumption. Analogously, in [25], the authors automatically generate a highly optimized stencil code for

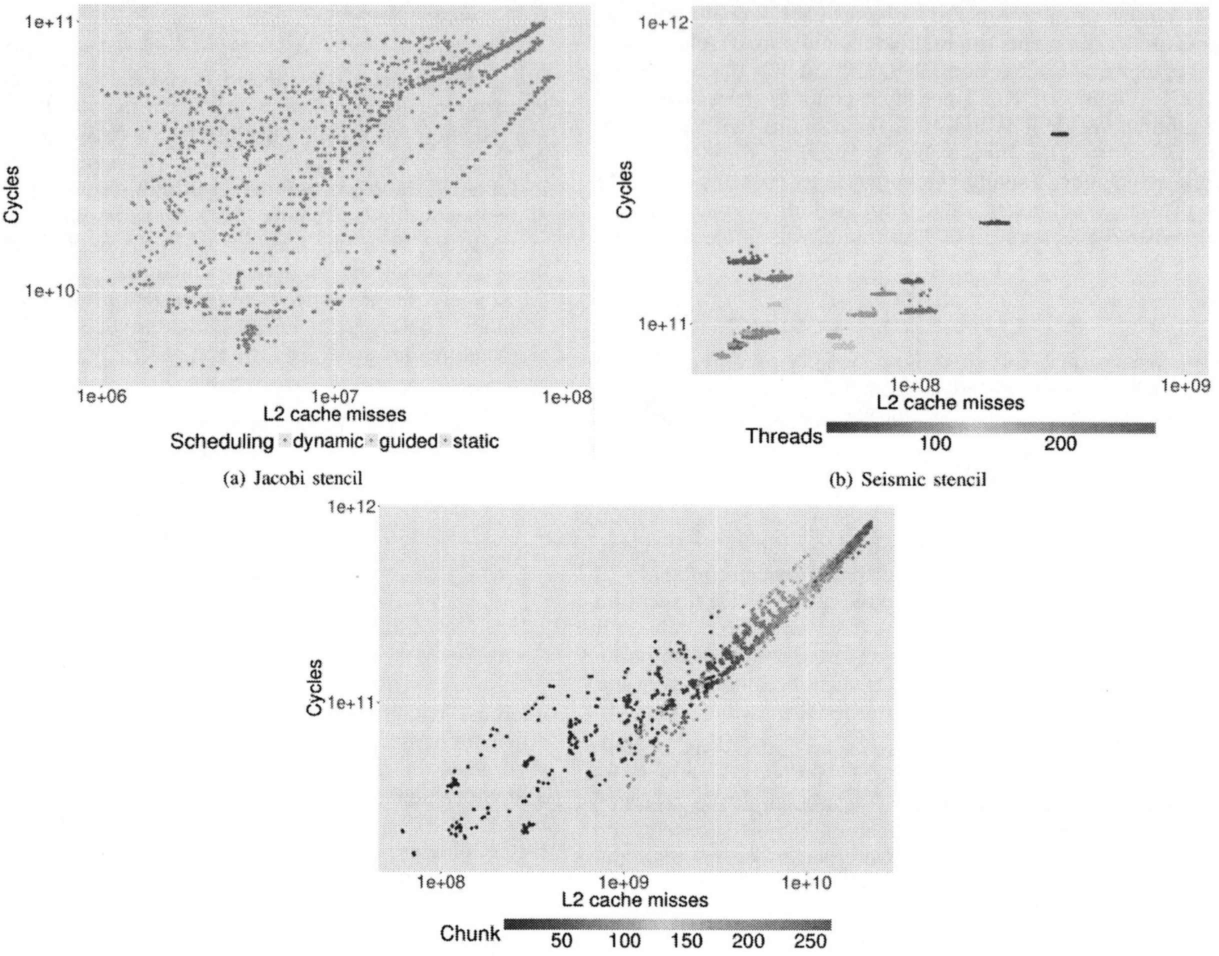

(a) Jacobi stencil

(b) Seismic stencil

(c) Acoustic stencil

Figure 6. Hardware counters behavior.

multiple target architectures. Overall, the main problem of these works is that the search domain can be very large and searching for the best configuration would take too much time.

Other works investigated the accuracy of regression models and ML algorithms in different contexts. In [26] the authors compared ML algorithms for characterizing the shared L2 cache behavior of programs on multi-core processors. The results showed that regression models trained on a given L2 cache architecture are reasonably transferable to other L2 cache architectures. In [27] the authors proposed a dynamic scheduling policy based on a regression model that is capable of responding to the changing behaviors of threads during execution.

Finally, in [28] the authors applied ML techniques to explore stencil configurations (code transformations, compiler flags, architectural features and optimization parameters). Their approach is able to select a suitable configuration that gives the best execution time and energy consumption. In [18], the authors improved performance of stencil computations by using a model based on cache misses. In [17], the authors proposed a ML model to predict and to optimize the performance of stencil computations on multi-core architectures.

VI. CONCLUSION

We presented a ML-based model to predict the performance of stencil computations on many-core architectures. We showed that the performance of three well-known stencil kernels (7-point Jacobi, seismic and acoustic wave propagation) can be predicted with a high accuracy by using the hardware counters.

Results from this work extend the ML-based strategy described in [17] for performance optimization of the elastodynamics equation on multi-core architectures. Our model is not restricted to Xeon Phi platforms and can also be implemented into architectures with the available hardware counters to obtain the cache-related measures. We used the PAPI library but we think that it could be implemented with another library (i.e., perf, pin, etc.). The future work can be summarized in two statements: First, we believe that our model can be integrated into an auto-tuning framework to find the best performance configuration for a given stencil kernel; second, we intend to design a model based on unsupervised ML algorithms to further improve our results.

Acknowledgments

This work has been granted by the *Coordenação de Aperfeiçoamento de Pessoal de Nível Superior (CAPES)*, the *Conselho Nacional de Desenvolvimento Científico e Tecnológico (CNPq)*, the *Fundação de Amparo à Pesquisa do Estado do Rio Grande do Sul (FAPERGS)*. Research has received funding from the EU H2020 Programme and from MCTI/RNP-Brazil under the *HPC4E Project*, grant agreement n.º 689772. It was also supported by Intel under the Modern Code project, and *PETROBRAS* oil company provided the acoustic wave numerical kernel code, under Ref. 2016/00133-9. The authors also thank to Bureau de Recherches Géologiques et Minières *(BRGM)* for providing the 7-point Jacobi and the seismic wave propagation codes.

References

[1] F. Dupros, H. Do, and H. Aochi, "On scalability issues of the elastodynamics equations on multicore platforms," in Proceedings of the International Conference on Computational Science, ICCS 2013, Barcelona, Spain, 5-7 June, 2013, 2013, pp. 1226–1234.

[2] V. Martínez et al., "Towards seismic wave modeling on heterogeneous many-core architectures using task-based runtime system," in Computer Architecture and High Performance Computing (SBAC-PAD), 2015 27th International Symposium on, Oct 2015, pp. 1–8.

[3] D. Carastan-Santos and R. Y. de Camargo, "Obtaining dynamic scheduling policies with simulation and machine learning," in Proceedings of the International Conference for High Performance Computing, Networking, Storage and Analysis, ser. SC '17. New York, NY, USA: ACM, 2017, pp. 32:1–32:13. [Online]. Available: http://doi.acm.org/10.1145/3126908.3126955

[4] Y. Li et al., "Capes: Unsupervised storage performance tuning using neural network-based deep reinforcement learning," in Proceedings of the International Conference for High Performance Computing, Networking, Storage and Analysis, ser. SC '17. New York, NY, USA: ACM, 2017, pp. 42:1–42:14. [Online]. Available: http://doi.acm.org/10.1145/3126908.3126951

[5] K. Datta et al., "Optimization and performance modeling of stencil computations on modern microprocessors," SIAM Rev., vol. 51, no. 1, Feb. 2009, pp. 129–159. [Online]. Available: http://dx.doi.org/10.1137/070693199

[6] ——, "Stencil computation optimization and auto-tuning on state-of-the-art multicore architectures," in Proceedings of the 2008 ACM/IEEE Conference on Supercomputing, ser. SC '08. Piscataway, NJ, USA: IEEE Press, 2008, pp. 4:1–4:12. [Online]. Available: http://dl.acm.org/citation.cfm?id=1413370.1413375

[7] A. Nguyen et al., "3.5-d blocking optimization for stencil computations on modern cpus and gpus," in 2010 ACM/IEEE International Conference for High Performance Computing, Networking, Storage and Analysis, Nov 2010, pp. 1–13.

[8] F. Dupros, F. Boulahya, H. Aochi, and P. Thierry, "Communication-avoiding seismic numerical kernels on multicore processors," in International Conference on High Performance Computing and Communications (HPCC), Aug 2015, pp. 330–335.

[9] D. Michéa and D. Komatitsch, "Accelerating a three-dimensional finite-difference wave propagation code using gpu graphics cards," Geophysical Journal International, vol. 182, no. 1, 2010, pp. 389–402. [Online]. Available: http://dx.doi.org/10.1111/j.1365-246X.2010.04616.x

[10] R. F. Vilela, "Profiling of the finite differences wave equation resolution problem in xeon phi coprocessor," Master's thesis, Federal University of Rio de Janeiro, COPPE, http://www.pee.ufrj.br/index.php/pt/producao-academica/dissertacoes-de-mestrado/2017/2016033170-53/file, 2017.

[11] S. H., "Intel® Xeon Phi™ Processor 7200 Family Memory Management Optimizations," https://software.intel.com/en-us/articles/intel-xeon-phi-processor-7200-family-memory-management-optimizations, 2016, [retrieved: april, 2018].

[12] M. P., "An Intro to MCDRAM High Bandwidth Memory) on Knights Landing," https://software.intel.com/en-us/blogs/2016/01/20/an-intro-to-mcdram-high-bandwidth-memory-on-knights-landing, 2016, [retrieved: april, 2018].

[13] Intel, "OpenMP* Loop Scheduling," https://software.intel.com/en-us/articles/openmp-loop-scheduling, 2014, [retrieved: april, 2018].

[14] P. Mucci and ICL, "PAPI Library," http://icl.cs.utk.edu/papi/, 2014, [retrieved: april, 2018].

[15] C. Cortes and V. Vapnik, "Support-vector networks," Machine Learning, vol. 20, no. 3, 1995, pp. 273–297. [Online]. Available: http://dx.doi.org/10.1007/BF00994018

[16] H. Drucker et al., "Support vector regression machines," in Advances in Neural Information Processing Systems 9. MIT Press, 1997, pp. 155–161.

[17] V. Martínez, F. Dupros, M. Castro, and P. Navaux, "Performance improvement of stencil computations for multi-core architectures based on machine learning," Procedia Computer Science, vol. 108, 2017, pp. 305 – 314, international Conference on Computational Science, {ICCS} 2017, 12-14 June 2017, Zurich, Switzerland.

[18] R. de la Cruz and M. Araya-Polo, Modeling Stencil Computations on Modern HPC Architectures. Cham: Springer International Publishing, 2015, pp. 149–171.

[19] K. Datta et al., Scientific Computing with Multicore and Accelerators. Taylor & Francis Group: CRC Press, 2010, ch. Auto-Tuning Stencil Computations on Multicore and Accelerators.

[20] C. Andreolli et al., "Chapter 23 - Characterization and Optimization Methodology Applied to Stencil Computations," in High Performance Parallelism Pearls, J. Reinders and J. Jeffers, Eds. Boston: Morgan Kaufmann, 2015, pp. 377 – 396.

[21] D. Caballero et al., "Optimizing Fully Anisotropic Elastic Propagation on Intel Xeon Phi Coprocessors," in 2nd EAGE Workshop on HPC for Upstream, 2015, p. 4.

[22] S. Kamil et al., "An auto-tuning framework for parallel multicore stencil computations," in IEEE International Symposium on Parallel Distributed Processing (IPDPS), April 2010, pp. 1–12.

[23] J. Cronsioe, B. Videau, and V. Marangozova-Martin, "BOAST: Bringing Optimization through Automatic Source-to-Source Transformations," in International Symposium on Embedded Multicore/Manycore System-on-Chip (MCSoC). Tokyo, Japan: IEEE Computer Society, 2013, pp. 129 – 134.

[24] X. Niu, Q. Jin, W. Luk, and S. Weston, "A Self-Aware Tuning and Self-Aware Evaluation Method for Finite-Difference Applications in Reconfigurable Systems," ACM Trans. on Reconf. Technology and Systems, vol. 7, no. 2, 2014, pp. 15:1–15:19.

[25] N. Kukreja et al., "Devito: Automated fast finite difference computation," in Procs. of the 6th Intl. Workshop on Domain-Spec. Lang. and High-Level Frameworks for HPC, ser. WOLFHPC '16. IEEE Press, 2016, pp. 11–19.

[26] J. K. Rai, A. Negi, R. Wankar, and K. D. Nayak, "On prediction accuracy of machine learning algorithms for characterizing shared l2 cache behavior of programs on multicore processors," in International Conference on Computational Intelligence, Communication Systems and Networks (CICSYN), July 2009, pp. 213–219.

[27] L. Weng, C. Liu, and J.-L. Gaudiot, "Scheduling optimization in multicore multithreaded microprocessors through dynamic modeling," in Proceedings of the ACM International Conference on Computing Frontiers, ser. CF '13. New York, NY, USA: ACM, 2013, pp. 5:1–5:10.

[28] A. Ganapathi, K. Datta, A. Fox, and D. Patterson, "A case for machine learning to optimize multicore performance," in Proceedings of the First USENIX Conference on Hot Topics in Parallelism, ser. HotPar'09. Berkeley, CA, USA: USENIX Association, 2009, pp. 1–1.

Energy Determination of Superconducting Vortex Lattices with Stochastic Methods Calculated on GPUs

Manuel Rodríguez-Pascual, José A. Moríñigo, Rafael Mayo-García
CIEMAT
Avda. Complutense, 40
28040 Madrid, Spain
email: {manuel.rodriguez, josea.morinigo, rafael.mayo}@ciemat.es

Abstract—**By fabricating Nb films on top of array of Ni nanodots with different geometries, the vortex lattice for specific values of the external applied magnetic field is modified by the array of periodic pinning potentials. In this work, a GPU-based code developed from scratch simulating this phenomenon is presented. It evaluates the vortex–vortex and the vortex–nanodot interactions providing the total interaction between vortices and pinning sites, as well as the position of the vortices in the array unit cell. This final position is obtained with different stochastic processes being able to simulate square, rectangular, or triangular arrays of nanodefects of different size. A computational performance study is also made.**

Keywords- Superconducting vortices; nanostructures; heuristics; supercomputation; GPU.

I. INTRODUCTION

Superconducting vortex lattice dynamics and vortex lattice pinning are strongly modified by arrays of nanodefects embedded in superconducting films. By using arrays of holes (antidotes), which thread the films or dots embedded in the sample, this effect can be studied. Thus, many effects can be observed on these hybrid samples, for example, reconfiguration of the vortex lattice, effects induced by arrays made with different materials, different diameters of the pinning centers, channeling effects, etc.

Magnetoresistance measurements are a perfect tool to study these effects, since resistance vs. applied magnetic fields shows deep minima when the vortex lattice matches the unit cell of the array due to geometric matching occurs when the vortex density is an integer multiple of the pinning center density. These phenomena are ruled by the balance among different interactions, (i) vortex–vortex, (ii) vortex–artificially induced pinning center (array of nanodefects), (iii) vortex–intrinsic and random pinning centers. The magnetoresistance minima show up only when the temperature is close to T_c, since, at these temperatures, the effect of the vortex–array interaction is enhanced. The large roughness of the sample surface precludes the use of standard local probe methods to detect experimentally the vortex position and symmetry of the vortex lattice, which could be only inferred from the experimental matching conditions.

Therefore, theoretical approaches have been undertaken by computer simulation methods. In a pioneering work, in the framework of molecular dynamics, Reichhardt et al. [1], by integrating numerically the Langevin equation of motion, were able to predict some of the matching fields at which commensurate vortex arrangements happen. According to this approach, the superconducting penetration depth is the crucial parameter. These authors use cut-off conditions, pinning strengths and other relevant parameters governed by the penetration superconducting depth. In a similar scenario, Langevin equation of motion of the vortices, Dinis et al. [2] have been able to simulate the rectifier behavior of the vortex lattice in the transverse ratchet effect. In this case, the parameters are taken from the experiment and the random intrinsic pinning of the superconductor plays a crucial role. Simulations of vortex dynamics in superconducting films with pinning array have been also reported by Kato and Enomoto [3], Gropp et al. [4] and Rodríguez-Pascual et al. [5].

In this work, the possibility to simulate the commensurability experiments in the framework of the Langevin equation of motion is explored without any initial conditions neither constraints and using only as input the vortex–vortex interaction and the periodic pinning sites (array unit cell). This process is carried out with genetic algorithms and simulated annealing techniques on GPUs. The experimental magnetoresistance minima permit obtaining the number of vortices in the array unit cell and figuring out the vortices position for different arrays and matching fields, as well as evaluating the vortex lattice interaction.

The article reads as follows. After this Introduction section, the experiment that can measures the energy of superconducting vortex lattices is described. In Section III, the simulation that has been implemented with a code that runs on GPUs is presented, as well as the stochastic methodologies that have been designed. The obtained results are included in Section IV, while the conclusions come in Section V.

II. EXPERIMENT

Superconducting/magnetic hybrids have been grown by magnetron sputtering, electron beam lithography and etching techniques, for more details see for example. In brief, the samples are Nb film on top of array of submicrometric Ni dots which have been fabricated by electron beam lithography on Si (100) substrates. Thus, 400×600 nm^2 rectangular arrays of Ni dots have been selected as the artificially fabricated pinning arrays for the present work, though the code can also simulate squares, rectangles, and triangles of any size. The thickness of the Ni dots is 40 nm, while the thickness of the Nb film is 100 nm. The diameter of the Ni dots is 200 nm. The maximum number of vortices that could accommodate one of these pinning sites, i.e., the so-called filling factor could be estimated as one vortex per dot.

A cross-shaped bridge of 40 μm is patterned in the hybrids for magnetotransport measurements by means of standard photolithography and etching techniques. The magnetic fields are applied perpendicular to the sample and magnetoresistance measurements have been done in a commercial cryostat with superconducting solenoid.

Minima appear at applied magnetic fields $H_n = n \cdot \varphi_0/(a \cdot b)$, where a and b are the lattice parameters of the rectangular array and $\varphi_0 = 2.07 \cdot 10^{-15}$ Wb is the fluxoid. The number of vortices n per array unit cell can be known by simple inspection of the magnetoresistance curves, in which the first minimum corresponds to one vortex per unit cell, the second minimum to two vortices per unit cell, and so on.

III. SIMULATIONS

The next step is to model these behaviors by computer simulation. This have been done by implementing the DiVoS code. Computer simulation with DiVoS reproduces the aforementioned experimental effects, but different geometries of lattices have been also evaluated by calculating the interaction of each possible vortices configuration and choosing the most convenient, i.e., the one with the lowest energy according to the desired specifications (physical parameters) used as input data. This code has been implemented from scratch; it does not take advantages neither of matching conditions with respect to the vortices lattices nor computational cutoff approximations to place the vortices. Several interactions are present and the code obtains the configuration with the lowest energy, so the interactions in the overdamped equation of vortex motion can be described as follows:

$$fi = f_i^{vv} + f_i^{vp} = \Sigma_{j=1}{}^{N}{}_{V} f_0 K_0(\mid r_i - r_j \mid / \lambda) + \\ + \Sigma_{k=1}{}^{N}{}_{p} (f_p/rp)|r_i - r_k| \, \Theta[(r_p - |r_i - r_k| / \lambda)]\mathbf{r}_{ik} \qquad (1)$$

where f_i is the total force per unit length acting on vortex i, f_i^{vv} is caused by the vortex–vortex interaction and f_i^{vp} is the pinning force.

The first sum runs up to the total number of vortices N_v and K_0 is the zero order modified Bessel function, which depends on the distance r_{ij} and the penetration depth λ, being λ (at 0.99T_c) = 2.6 μm in the experiment. Specifically, f_0 is $3.08 \cdot 10^{-6}$ T^2nm in our experiment.

In addition, the second sum related to pinning force has k as the index referring to the different pinning sites in the system, Θ as the Heaviside step function, f_p as the maximum pinning force (it has been considered as 0.5 times the constant f_0) and r_p as the pinning radius (100 nm in our experiment).

The DiVoS code represents the following physical model: Surface is represented as a 2D grid; pinning sites define either rectangular or triangular cells; there are 1,2...n vortices per cell; and, vortex-vortex and vortex-pinning site interactions rule the system according to the previous formula (1).

As for the previous experiments, the number of cells sums up to 60x60. Considering each cell could contain up to 3 vortices, for example, the problem to be tackled results in 7,200 vortices. The interaction of 2 vortices is simulated by calculating the distance first and applying the Bessel Modified Function afterwards. Doing so, the vortices dynamics is performed as a vortex in a given position moves by looking at the interactions with all the others and the pinning sites and moving to the less energetic adjacent position. Thus, there are about 25 million interactions to be calculated in every simulation step; in other words, considering for example a rectangular cell size of 400x600 nm (simulated points), there are 240,000 positions for each vortex.

Altogether, an efficient way of calculating the system energy and algorithms to discard most of the possible configurations is needed, which results in stochastic processes such as Genetic Algorithm and/or Simulated Annealing in this work. Stochastic processes have demonstrated their correct approach [6].

A. *Genetic algorithm*

A genetic algorithm has been implemented as part of the code. As it is well known, a population of candidate solutions (called genes) to an optimization problem is evolved toward better solutions. Each candidate solution has a set of properties which can be mutated and altered randomly as the process goes on.

The evolution starts from a population of randomly generated individuals, and is an iterative process. In each generation, the fitness of every individual in the population is evaluated; the fitness is usually the value of the objective function in the optimization problem being solved, i.e. previous formula (1). The more fit individuals are stochastically selected from the current population, and each individual's genome is modified (recombined and possibly randomly mutated) to form a new generation. The new generation of candidate solutions is then used in the next iteration of the algorithm. The algorithm terminates when either a maximum number of generations has been produced, or a satisfactory fitness level has been reached for the population.

Once the genetic representation and the fitness function are defined, the genetic algorithm proceeds to initialize a population of solutions and then to improve it through repetitive application of the mutation, crossover, inversion, and selection operators. The code initializes several genes and let them evolve in a way in which also mutation, crossover, inversion, and selection operators can do cross-fitting.

B. Simulatted annealing

Simulated annealing uses a probabilistic technique for approximating the global optimum of the given function (1). It is a metaheuristic to approximate global optimization in a large search space, mainly used when the search space is discrete as it is the case. Simulated annealing is usually preferable to alternatives such as gradient descent for problems where finding an approximate global optimum is more important than finding a precise local optimum in a fixed amount of time.

In the implemented code, f_i moves to f_{i+1} via a proposal, i.e. the vortices move randomly to any of the 8 adjacent positions (N, NE, E, SE, S, SW, W, and NW). If the new state has lower energy, then f_{i+1} is accepted; unlike, f_{i+1} is accepted with probability $A=exp(-\Delta f/KT)$ By doing so, stochastic acceptance of higher energy states allows the process to escape local minima. If the vortices move 1 by 1 and the comparison is made, the code reproduces molecular dynamics processes; if all the vortices move at the same time in the single step, multidimensional Gaussian is carried out. When T is high, the acceptance of these moves is higher, and local minima are discouraged. As T is lowered, more concentrated search near current local minima is performed due to only few moves will be allowed. Thus, if we get the temperature decrease schedule right, it can be hoping that there will be converge to a global minimum.

Reannealing interval, or epoch length (L), is the number of points to accept before reannealing (change the temperature), i.e., L represents the number of iterations at a particular temperature. Larger decreases in T require correspondingly longer L to re-equilibrate. Also, running long L at larger temperatures is not very useful, so T is decreased rapidly at first. Reannealing interval evolves with $L_{k+1}=\beta L_k$ with $\beta>1$.

Thermostat can be simulated in three different ways:

- Linear: Temperature decreases as $T_{k+1}=\alpha T_k$, (with $1<\alpha<0$) or $Tk-\alpha$ (with $\alpha>0$)
- Exponential: Temperature decreases as 0.95^{α} with $\alpha \geqslant 1$
- Logarithmic: Temperature decreases as $1/\log(\alpha)$ with $\alpha \geqslant 10$

IV. Results

In this work, preliminary results of the execution of the code for values of the matching point 1, 2, and 3 are reported. They are simply intended to test the computational performance as well as the behavior of the different stochastic processes.

Regarding the parallelization, the GPU cluster located at CETA was used. Out of the whole amount of resources, several GPU cards were used so the number of Nvidia cores moved between 1,000 and 5,000. Also, several outcomes have been cloncluded:

- Over 99% of the computing time gets into the evaluation of interactions
- Original scalability is of $O(N^2)$, being N the number of vortices
- Parallel scalability is of $O(N^2/2G)$, being G the number of GPU cores
- By using a cache mechanism, a GPU core is faster than a CPU for this problem

With respect to the genetic algorithm, up to 200 individuals were randomly created and allocated in different Nvidia cores in order to be executed independently. This test performed up to 500 generations, i.e., number of steps to allow the individual to evolve. Mutation and crossover rate are constant

The data related to the simulated annealing were as follow:

- Initial value of T was 25,000 K
- Temperature decreased as $T_{k+1} = 0.8*T_k$
- Population in the simulated annealing is 1

Also, an experiment which combined simulated Annealing and Genetic experiment was carried out. It was programmed in a way in which the output of a simulated annealing (20 epochs) was employed as input for the genetic algorithm (population 20).

All experiments have been executed 5 times and the results are the average values. In order to properly depict the obtained results with figures as large as possible, the latter are shown below. As it can be seen in Figs. 1 and 2, that for the genetic algorithm case, there is no influence of the number of individuals in the obtained result of the energy of the system as it is roughly constant. Unlike, the higher the number of generations, the better result.

Regarding the execution time, it increases as the number of generations increases linearly, but it shows a worse behavior with the number of individuals generated. Thus, there is a double reason for not increasing the population unnecessarily.

The simulated annealing case is more interesting (see Fig.3). It provides the best result both in execution time and energy. Also, it can be deducted that the ideal number of epochs is close to 100 as it is the region where the energy decreases significantly. But not only due to the physical interest, which is the higher one in order to obtain valid results, the execution time vs. the number of epochs also demonstrate that from 150 on the system saturates.

With respect to the hybrid case, it does not provide a significant enhancement as the genetic algorithm behavior penalize the simulated annealing one.

V. Conclusions

Hybrid superconducting/magnetic samples are fabricated with superconducting films on top of array of pinning centers. The magnetoresistance of these hybrids, close to critical temperature, shows deep and equal spaced minima which are due to commensurability effects between the vortex lattice and the unit cell of the array. The first minimum appears when the density of the pinning centers equals the density of the vortex lattice, upper order minima take place at $H_n = n(\Phi_0/S)$, where $n>1$ is an integer number and Φ_0 is the quantum fluxoid. Taking into account the vortex–vortex and the vortex–pinning center interactions, a GPU-based computing simulation code has been implemented. This code can calculate different values and positions for different lattices in size, matching field values, and geometry of the pinning sites, which allows having a picture of the different vortex lattices.

The position of the vortices, as well as the minimum energy are obtained with three different stochastic methodologies: genetic algorithm, simulated annealing, and, a combination of the former. The best results are provided by the simulated annealing version.

Further tests are needed to be carried out in order to properly match experiment and simulation switching different parameters for the simulated annealing case.

ACKNOWLEDGMENT

This work was partially funded by the Spanish Ministry of Economy, Industry and Competitiveness project CODEC2 (TIN2015-63562-R) with European Regional Development Fund (ERDF) and by the European Commission H2020 project HPC4E (Grant Agreement n 689772). This work was carried out on the computing facilities provided by the Extremadura Research Centre for Advanced Technologies (CETA-CIEMAT) with ERDF funds and the CYTED Network RICAP (517RT0529).

REFERENCES

[1] C. Reichhardt, C.J. Olson, F. Nori, "Commensurate and Incommensurate Vortex States in Superconductors with Periodic Pinning Arrays", Phys. Rev. B 57(13), pp. 7937-7943, 1998.

[2] L. Dinis, D. Pérez de Lara, E.M. González, J.V. Anguita, J.M. Parrondo, J.L. Vicent, "Transverse Ratchet Effect and Superconducting Vortices: Simulation and Experiments", New J. Phys. 11, 073046, 2009.

[3] R. Kato and Y. Enomoto, "Simulations of Vortex Dynamics in Superconducting Films with Pinning Array". Physica B: Condensed Matter 284–288, pp. 899-900, 2000.

[4] W.D. Gropp, H.G. Kaper, G.K. Leaf, D.M. Levine, M. Palumbo, V.M. Vinokur, "Numerical Simulation of Vortex Dynamics in Type-II Superconductors", J. Computational Physics 123(2), pp.254-266,1996.

[5] M. Rodríguez-Pascual, A. Gómez, R. Mayo-García, D. Pérez de Lara, E.M. González, A.J. Rubio-Montero, J.L. Vicent, "Superconducting Vortex Lattice Configurations on Periodic Potentials: Simulation and Experiment", J. Supercond. 25, pp. 2127-2130, 2012.

[6] N.G. Van Kampen, "Stochastic Processes in Physics and Chemistry", 3ed., North Holland, 2007.

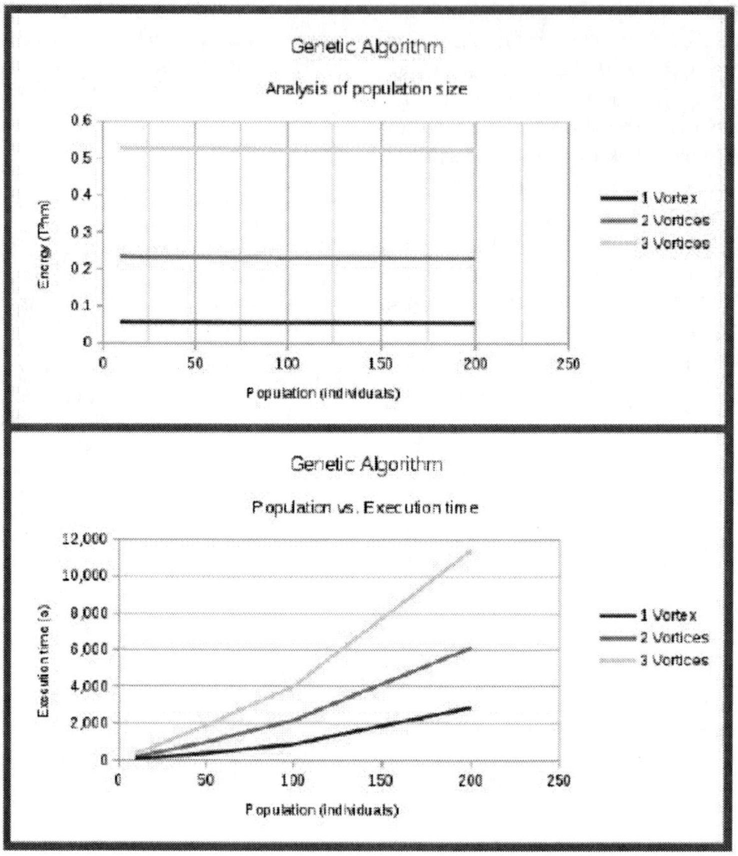

Figure 1. Genetic algorithm version of the code. Influence of the population on the energy obtained and the executiont time.

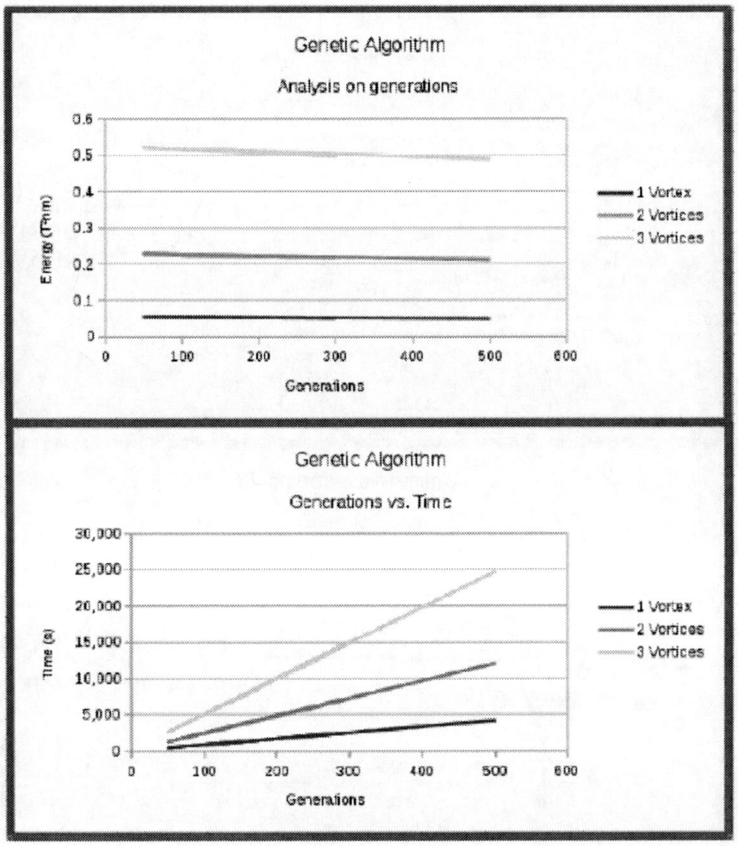

Figure 2. Genetic algorithm version of the code. Influence of the generations on the energy obtained and the executiont time

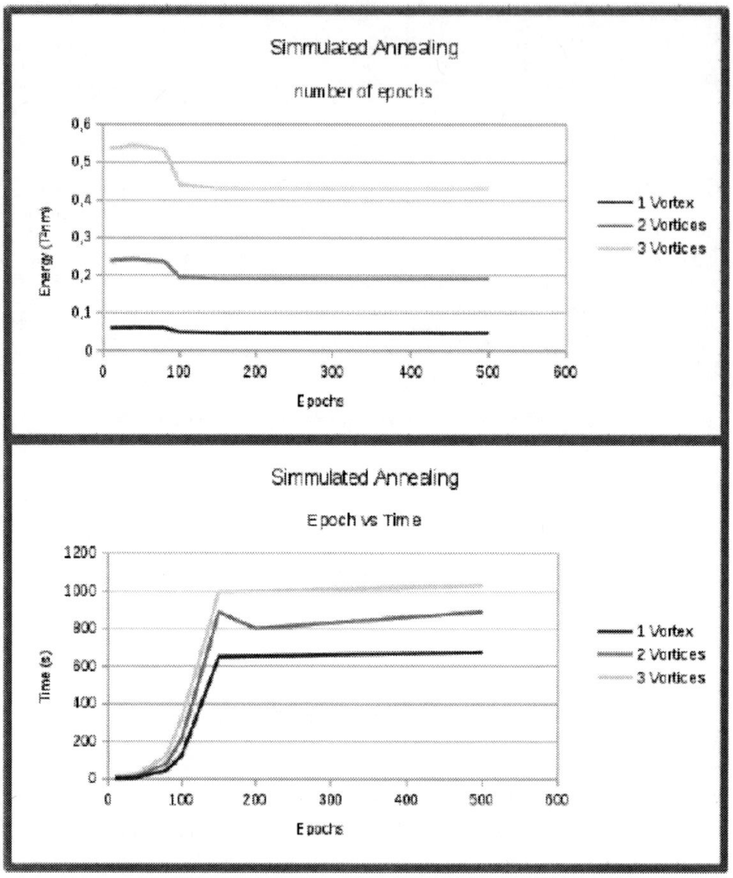

Figure 3. Simulated annealing case. Influence of the number of epochs in the obtained energy amdon the execution time